D0833835

700039161858

How many friends does one person need?

by the same author

Grooming, Gossip and the Evolution of Language
The Trouble With Science
The Human Story

Robin Dunbar

How Many Friends Does One Person Need?

Dunbar's Number and other evolutionary quirks

faber and faber

First published in 2010
by Faber and Faber Limited
Bloomsbury House, 74–77 Great Russell Street,
London WC1B 3DA

Typeset by Faber and Faber
Printed in England by CPI Mackays, Chatham, Kent

All rights reserved
© Robin Dunbar, 2010

The right of Robin Dunbar to be identified as author of this work has been asserted in accordance with Section 77 of the Copyright, Designs and Patents Act 1988

A CIP record for this book
is available from the British Library

ISBN 978–0–571–25342–5

2 4 6 8 10 9 7 5 3 1

Contents

Acknowledgements

This volume had its origins in a series of popular science articles that I wrote for *New Scientist* magazine (mostly between 1994 and 2006) and the *Scotsman* newspaper (between 2005 and 2008). In bringing them together in this volume, my intention has been to give some flavour of the excitement – and some of the fun – that has pervaded the evolutionary study of behaviour, and in particular human behaviour, over the last decade. I am grateful to both for providing me with an opportunity to indulge a passion for popular science writing over the years, as well as for allowing me to reuse these pieces in this volume. I also thank the *Observer*, *Scotland on Sunday*, the *Times Higher Education Supplement*, the Royal College of Physicians (London), Charles Pasternak and OneWorld Books, and Faber and Faber for permission to reuse individual pieces published by them. Most of these pieces have been substantially edited or adapted for this volume.

Pieces published in the *Scotsman* make up the bulk of chapters 2, 4, 5, 8, 9, 10, 12, 13 and 16, and feature in chapters 3, 6, 11, 14, 17, 19 and 21. Pieces published in *New Scientist* magazine appear in chapters 7, 13, 14 and 21, and make up the bulk of chapters 3, 17, 18, 20 and

22. A piece published in the *Observer* contributes to chapter 7, and one from *Scotland on Sunday* to chapter 21. An article from the *Times Higher Education Supplement* makes up the bulk of chapter 15. Part of chapter 3 appeared in *The Science of Morality* (2007; edited by G. Walker, published by the Royal College of Physicians, London); part of chapter 12 originally appeared in my *The Human Story* (2004, Faber and Faber); and part of chapter 15 appeared in *What Makes Us Human* (2007; edited by Charles Pasternak, published by OneWorld Books, Oxford).

Finally, I am grateful to my agent Sheila Ableman, and to my editor at Faber, Julian Loose.

Chapter 1

In the Beginning

We share a history, you and I. A history in which our respective stories snake back through time, edging ever closer to each other until finally they meet up in a common ancestor. Perhaps our lineages meet up only a few generations back, or maybe it was a thousand years ago. Perhaps it was so long ago that it predates history – though even that could not have been more than two hundred thousand years ago, a mere twinkle in earth time. For we modern humans all descend from a common ancestor who roamed the plains of Africa a mere ten thousand generations ago, ten thousand mothers giving birth to ten thousand daughters . . . no more than would fit in a town of very modest size today.

For us, that has two important implications. One is that we share most of our traits in common. From Alaska to Tasmania, and Tierra del Fuego to Spitzbergen, we are a single family, one biological species united by common ancestry. The other is that those traits we share are, nonetheless, the product of evolution, honed by the demands of the lives that our ancestors led. Sometimes, they are the product of deep evolutionary time, traits we share with the other members of our biological family, the great apes, and especially the African great apes. Sometimes, those traits are of more recent origin, wrought in the fire of the particular circumstances that our more immediate ancestors faced in the battle for life, traits that mark us out as human – not special, because we are just one of many tens of thousands of individually unique species of animals, but unique in that we alone possess them. Some of these give us the capacity for culture, that remarkable product of the human mind that has made us what we are – those traits that allowed us to break away from our biological roots, that allowed human history to be what it is.

Yet, in our enthusiasm for the wonders of human culture, we sometimes overlook just how much of our behaviour is rooted in our biological evolution. The human mind is surely one of the wonders of the natural world, yet sometimes it seems so pedestrian and constrained that it is hard to see how we differ from any of the other primates. We live in massive conurbations numbering tens of millions of individuals, a product of our cultural flexibility if ever there was one. We have lived in villages only for the last ten thousand years, and cities the size of Bombay or Rio de Janeiro only for the last century at most. These are novel innovations, a product of our capacity to invent ways of making do. Yet, at the same time, our social world is still what it was several hundred thousand years ago. The number of people we know personally, whom we can trust, whom we feel some emotional affinity for, is no more than 150, Dunbar's Number. It has been 150 for as long as we have been a species. And it is 150 because our minds lack the capacity to make it any larger. We are as much the product of our evolutionary history as any other species is.

I probably owe my interest in evolution to my American grandmother. Though a fiercely God-fearing Presbyterian missionary, she was also a surgeon and sufficiently well-versed in science to be an enthusiast for the new discoveries in human evolution that were emerging from Africa during the 1950s. When I was ten or eleven, she sent me a series of Audubon Society booklets on every imaginable subject to do with the natural world, complete with sticky stamps to paste in. One was on evolution, and covered everything from dinosaurs to humans. I became hooked on the story of human evolution. Some years later, I read Darwin's *Origin of Species*, having found it by chance in the school library. It was interesting, but I can't say I got a great deal out of it at the time. I was becoming more interested in philosophy, and science wasn't really my thing.

Then, five or six years later as a postgraduate student, I was thrust willy-nilly back into Darwin's world. I was deeply engaged in studying the behaviour of monkeys in the wild, spending several years doing fieldwork in Africa during the early 1970s. At the time, evolutionary thinking in the behavioural sciences was apt to be somewhat loose and wayward. We returned from fieldwork in Ethiopia in late 1975 to find the world had been turned upside down. Edward O. Wilson had just published his *Sociobiology: The New Synthesis* and Richard Dawkins would publish *The Selfish Gene* the following year. It was a life-changing experience for all of us. Overnight, we were made to think about evolutionary processes in a much more rigorous way. We were being asked to return to a more strictly Darwinian view, after decades of increasingly lax, often speculative, thinking that had come to characterise much of organismic biology in mid-century. Of course, neither book invented something that was novel. What both, in their different ways, did was to lay out in stark detail the ideas that evolutionary biologists had slowly been developing over the previous decades.

The big intellectual change was a shift away from thinking that evolution was for the benefit of the species to one in which evolution was for the benefit of the genes that underpinned a trait, whether that trait was physical or behavioural. This should not be taken to imply that behaviour is hardwired, determined by the genes you inherit. Few traits are ever that simple in biology. But taking a gene's-eye view in which the benefits of a trait are costed out in terms of the impact they have on how often a particular gene is represented in the next generation brings us closer to Darwin's original conception of the theory of evolution by natural selection. More importantly, perhaps, it moved us away from the naïve genes-determine-all-behaviour view that has so often bedevilled thinking in this area to one in which an individual's freely made decisions on how to behave, free of any direct genetic input, could still be understood in a Darwinian framework. The following decades saw a veritable explosion of research. We learned so much in so short a space of time. Looking back, it is difficult now to convey the excitement of the time. So much of what was then novel is now accepted as fact.

Charles Darwin did not, of course, invent the theory of evolution. It had already had a long history within European biology dating back at least a century before young Charles was even a twinkle in his mother's eye. In fact, his own polymath of a grandfather, Erasmus Darwin, had himself made a seminal contribution to promoting the idea of evolution in one of his own best sellers. If anyone deserves the credit for inventing the theory of evolution it should probably be the great eighteenth-century French biologists – Cuvier, Buffon, Lamarck, among others. But they had been locked into a medieval mindset that had its origins in the views of Aristotle and Plato, filtered through the intellectual spectacles of the Church Fathers, a seminal group of medieval Christian theologians who established the core tenets of modern Christian theology. Building on the thinking of their Greek predecessors, they saw evolution as progressive, with each species inexorably climbing slowly but surely up the 'Great Chain of Being' from primitive life forms to join the angels just below God, who, at least as far as they were concerned, inevitably stood at the pinnacle of it all.

The publication of Darwin's book *On the Origin of Species* in 1859 set aside the old *scala natura*, or Great Chain of Being, that had been the linchpin of evolutionary thinking ever since Plato. Darwin set in train a new way of thinking about the natural world, a world whose history is driven by the demands of successful biological reproduction. In the process, of course, he upset quite a few apple carts, not least because his new vision of evolution challenged Victorian beliefs about the established order. Not only were Englishmen not the high point of evolution, but there wasn't that much room at the top for God either.

Darwin's great genius was to recognise that natural selection is the engine that drives evolution. In doing so, he dragged the theory of evolution out of the medieval doldrums into the modern world. He provided a mechanism that could explain how life on earth could have evolved without need for a creator. And it was a mechanism that, at the same time, could explain how and why a species might have evolved particular traits, traits that enabled individual animals to reproduce more successfully.

As with all scientific ideas, Darwin's theory underwent extensive development in the decades after the publication of the *Origin*. He expanded his ideas on natural selection to include sexual selection (selection for traits that enhance attractiveness to prospective mates). He applied his ideas to the nascent discipline of psychology – commenting at length on topics such as music, language, emotions and physical attractiveness – and even finally the evolution of Man.

Nor did his theory come to a halt with his death in 1882. It continued to be developed by those who came after him. We know so much more now than Darwin himself ever did, but the core of modern evolutionary theory and its many intellectual derivatives still lies firmly in Darwin's elegantly simple idea: organisms behave in ways that tend to enhance the frequencies with which the genes they carry are passed on to future generations.

It was into this heady atmosphere that I was thrust as a young researcher in the 1970s. We were galvanised and excited by the opportunities on offer, by the heady mix of new Darwinian theories whose strong predictions could guide our research and give us new questions we could ask that no one had thought of asking before. Looking back on three decades or so of this research is to realise what a privileged generation we had been. We witnessed a genuine scientific revolution as it happened. Our ways of thinking were changed for ever, just as the Victorians had had their worldview changed by Darwin. New conceptions of how animals behaved and evolved emerged that challenged our long-held assumptions about how the world was. A decade or so later, we began to apply these same ideas to human behaviour.

In the chapters that follow, I try to convey some of that excitement. Much of the research I will talk about is my own, or was done by members of my research group. But some of it will draw, somewhat idiosyncratically no doubt, on research by others that bears on the topics that have driven my own research over the past decade – why humans behave as they do, what it is to be human.

So, let me now invite you to explore with me those parts of you that, in the words of the advertisement, even the most proverbially exotic beers can never reach – how many friends you have, whether you have your father's brain or your mother's, whether morning sickness might actually be good for you (or, at least, for your baby), why Barack Obama's victory in the 2008 US presidential campaign was a foregone conclusion, why Shakespeare really was a genius, what Gaelic has to do with frankincense, and why we laugh. In the process, we'll examine the role of religion in human evolution, the fact that most of us have unexpectedly famous ancestors, and the reason why men and women never seem able to see eye to eye about colours. I'll couch all this in terms of evolution and Darwin's great insights, something that will make us ponder the very bases of science itself. But let's begin with the very core of what makes us human . . . our big brains.

Chapter 2

The Monogamous Brain

Of all the traits that natural selection has managed to evolve for us, our brains are surely the most valuable. Brains are the greatest evolutionary invention of all time. They were designed to free us from the worst of the evolutionary grind to which the rest of brute nature is subjected by allowing us to fine-tune our behaviour to circumstances. We can consider the options, weigh up the pros and cons, worry about the implications of behaving one way or the other, and then choose what seems like the most sensible thing to do. Thus it is that we rise above brute nature – a paragon of evolution. Or, at least, so it seems. In reality, brains are more complex than you might think. Yet, they are not quite as flexible and omniscient as we would like them to be. And we owe a good deal more of our brains to the vagaries of evolutionary history than we might wish.

Romeo, Romeo, wherefore art thou . . .?

Our brains are massively expensive, consuming about twenty per cent of our total energy intake even though they only account for about two per cent of our total body weight. That's a massive cost to bear, so brains really need to be spectacularly useful if they are going to be worth the cost. The consensus, at least for the primate family, is that we have our big brains to enable us to cope with the complexities of our social world. However, that story has recently acquired an interesting new twist as a result of studies on birds and other groups of mammals that my colleague Susanne Shultz and I have done. It seems that it is pairbonding that is the real drain on the brain. So let me ask: have you been struggling yet again with your partner's foibles? If you find relationships really hard work, then it seems you are in very good company. Among the birds and mammals in general, the species with the biggest brains relative to body size are precisely those that mate monogamously. Those that live in large anonymous flocks or herds and mate promiscuously have much smaller brains.

The birds make it especially clear that the real issue is strong, resilient, long-lasting pairbonds. Birds that mate monogamously come in two quite different kinds. There are those, like many common garden birds such as robins and tits, that choose a new mate each breeding season. But there are many others, such as many birds of prey, the owls and most of the crow and parrot families, that mate for life. It is this second group which have the biggest brains of all among the birds, far bigger than those that are seasonally monogamous, and this is true even when we control for differences in lifestyle, diet, and body size.

Among mammals, monogamy is much rarer (only about five per cent of mammals mate monogamously), but here too those that do so – including the many species of the dog/wolf/fox family, and antelope like the little klipspringer and the diminutive dikdik – have bigger brains than those that live in larger social groups where mating is promiscuous.

Biologists probably wouldn't get so excited about having a big brain, were it not for the fact that brain tissue is extremely expensive to grow and maintain – only your heart, liver and guts are more expensive. Evolving a bigger brain is thus no idle matter in evolutionary terms. And, given what brains do, this suggests that something about pairbonded relationships is significantly more taxing than life in the large anonymous flocks of shorebirds or the herds of deer and plains antelope. So what makes monogamous pairbonds so cognitively demanding?

One likely reason is that lifelong monogamy carries enormous risks. A poor choice of mate – one who is infertile, a lazy parent or prone to infidelity – risks jeopardising your contribution to the species' gene pool. Since, biologically speaking, that is what life is all about, it is not difficult to see that there are enormous evolutionary advantages to paying the cost of having a brain big enough to enable you to recognise the signs of a bad prospect when you see one. That way, you get to avoid a whole lot of trouble, and do better for yourself in the evolutionary stakes.

But there is another aspect to monogamy that might be just as important, and that's your ability to co-ordinate your behaviour with that of your mate. Consider the case of the average songbird in your garden. The business of mate choice is over, the female has laid her eggs, and now comes the tough bit – the long job of sitting on the nest while the eggs incubate, and the feeding of the fledglings that follow. Now, were it the case that one or other of the pair spent the whole of its day down at the avian equivalent of the pub, its mate would soon end up with the invidious choice between abandoning the eggs to cooling and predation so that it can feed, or staying on the nest and starving. For a small bird that has to eat its own body weight in food each day just to stay alive, this is no mean issue. In short, you need a mate that is smart enough to figure out what your needs are, and when it should return and take over its share of the nesting duties.

So perhaps it's the need to be able to factor your mate's perspective in to your own that is so cognitively demanding. Our own experiences would tell us that keeping a relationship on course through the years is a very delicate business, requiring a lot of fancy footwork to anticipate and see off at the pass all those potential sources of disagreement. Or, when they come from left field and we don't see them until they hit us, it's being able to see how to mend the fences and restore the equilibrium once again.

So as you struggle to figure out why your spouse has behaved so badly yet again, console yourself with the thought that evolution has blessed you with one of its crowning glories – a brain capable of figuring out how to get the best out of a bad job. After that, it's all plain sailing. Even the humble birds on your garden table can sort that one out.

Whose brain is it anyway?

Think about it: you have two parents, who each provide you with one set of genes, a complete set for everything about you. But you aren't just a fifty–fifty mixture of each of them. In most traits, you tend to resemble one or the other, so that by and large you end up as a kind of mosaic – your mother's nose, your father's chin, perhaps even your grandfather's hair through some quirk of a throwback to earlier generations. All this is pretty well understood, thanks mainly to the pioneering efforts in the 1850s of that indefatigable scientist-monk, Gregor Mendel, the founding father of modern genetics.

Now, one might expect that you would be a random mosaic of bits inherited from your two parents, and that these would vary between individuals – half the population would inherit a particular trait from their fathers, and the rest would inherit it from their mothers. It seems not. Instead, it turns out that some bits are always inherited from your mother and other bits always inherited from your father. The genes seem to know where they have come from, and which of them should switch themselves off (be 'silent' in the technical jargon).

The surprise is what happens in your brain. In an experimental study of natural genetic deficits in rats, Barry Keverne and his colleagues at Cambridge University found that animals with no maternal chromosomes lacked a fully developed neocortex, whereas those with no paternal chromosomes lacked a fully developed limbic system. This process whereby one set of genes is always 'silenced' is known as 'genomic imprinting'. Although the mechanisms involved are not yet fully understood, it seems that, in effect, individual genes 'know' whether they were paternal or maternal genes.

This finding gels rather neatly with another recent study. Rob Barton from Durham University and his colleagues have shown that, across the broad range of primate species, the size of a species' neocortex correlates best with the number of females in the group, whereas the size of the limbic system (part of the emotional response mechanism) correlates better with the number of males in the group. Since the number of females that a species can sustain in a typical group mainly reflects the females' social skills, this makes sense because the neocortex is related to social skills. On the other hand, in most primate species, male–male relationships are based more on competition for dominance rank (which is what allows males to be successful in the mating game), and this understandably has much more to do with males' willingness to fight.

The fact that the genomic imprinting is this particular way around is intriguing. In most primate species, the key to a female's reproductive success is the support she elicits from the sisterhood. For females to make their social relationships work, they need to be able to negotiate their way through a complex social world. Analysis of more than three decades of family histories from the population of baboons in Kenya's Amboseli National Park has shown that the females who are socially most successful also have the largest number of surviving offspring at the end of their lifetime.

But for males, the issue is much less about social skills than about willingness to keep slugging it out in a fight. Now, any sensible individual who gets involved in a fight will quickly realise that discretion is invariably the better part of valour and retire gracefully to live (and maybe fight) another day. But in the mating game, those who retire from the fray don't get the girl. So a mechanism that stops males thinking too much and lets the red mist take over usually works better. There may be a risk of injury or even death, but in a winner-takes-all game there is no point in being second. A small neocortex and a big limbic system is just what you want. If you have to fight for a living, best to bite first and think afterwards.

In effect, the females have won the battle over who controls the neocortex because social skills are more valuable to them, whereas males have won the battle over who controls the limbic system because it pays not to think too much about what you are doing if you get into a fight. The evolutionary battle of the sexes ends up being about control over the bits of the brain, though it is still something of a mystery as to how this is brought about. On second thoughts, I'm not so sure that I like the drift of this conversation . . . Perhaps we'll change the subject.

Four eyes better than three

Did you know that our eyes are actually part of our brain? They are an outgrowth of the brain that developed a sensitivity to light, came to the surface and, in doing so, allow us to see what's going on out there in the external world in a way that touch and smell cannot. As those who go blind through old age or accident know only too well, our life is ruled by vision – and especially the wonders of colour vision.

So, let me for a moment speak confidentially to the men. Have you, I wonder, become exasperated by your wife's fussing that the colours of her outfit clash when they seem perfectly fine to you? Well, she may be right: it seems that about a third of women see the world in four basic colours, whereas men only have the standard three (red, blue and green). These tetrachromatic (four-colour) women have an extra shade of green or an extra shade of red. Heaven forfend – some even have all five colours. It seems that some women really do see a very different world from the rest of us.

According to the standard story that they told us in school biology classes, we have two kinds of vision cells in the retina (the light-sensitive layer at the back of our eyeballs): rods give us the black-and-white vision that we use at night, and the cones give us colour that we use by day. The conventional wisdom is that there are three kinds of cones, each sensitive to a slightly different wavelength of light. These are red, blue and green, just as they are in the screen of your TV. We perceive the colours of the rainbow by the way the intensities of these three colours mix.

Now, the genes for two of these colours (the red–green dimension) are on the X chromosome, and those for blue are elsewhere, on chromosome seven. And this explains why men – but only very rarely women – are sometimes colour-blind and why this is usually red-blindness and almost never blue-blindness. Men only have one X chromosome (inherited from their mother), and if that chromosome is a bit dodgy, they don't have a back-up for any of the genes that are on it. Since women have two X chromosomes (one inherited from each parent), they always have a back-up in case of emergencies.

And this provides us with a very simple explanation for the four- (or five-) colour effect. Slight mutations of the genes that code for the colour-sensitive pigments in the retina can mean that different people see slightly different shades of red or green. For men, whatever shade you get from your single X chromosome is what you get: that's how you see the world. But women can end up with two slightly different shades of red or green on their two X chromosomes. If both X chromosomes become active during the development of the eyes, these women can have cones that code for both pigment sensitivities, and so end up with an extra colour dimension, in some cases even two extra ones – blue, red, shifted red, green and shifted green, five colours in all.

Now, here's where the tricky bit comes in. All this would be fine, because it would just mean that women live in a richer colour world than men, and who cares about that? But Mark Changizi and his colleagues at the California Institute of Technology in Pasadena now have an uncomfortable twist on this. Sex differences in colour sensitivity of this kind are far from unknown in primates: one particularly well-known one is the fact that, among the New World monkeys, females are trichromats (they have three-colour vision) but males see only two colours. Changizi and his colleagues noticed that sex difference in colour sensitivity in primates correlates with the amount of bare facial skin that a species has. Species which have large areas of bare skin that can change colour as a result of increased or decreased blood flow are precisely those that have full three-colour vision. They make the obvious connection: is the fact that humans are a 'naked ape' related to our good colour vision?

And here is where the salt gets rubbed into the wound. Perhaps women's sensitivity to colour (and especially reds) has something to do with their apparently mysterious capacity to know exactly when your protestations about where you have been all evening are, shall we say, just a little liberal with the truth. In short, do women know when men are lying because they can pick up much finer shades of blushing than their partners think they are giving away? How unkind can evolution possibly be?

Chapter 3

Dunbar's Number

The big social revolution of the last few years has not been some great political event, but the way our social world has been redefined by social networking sites like Facebook, MySpace and Bebo. Darwin and his contemporaries could not have conceived of such things, even in their wildest dreams. For a privileged few like Darwin himself, the geographical scatter of their friends might have been greatly enlarged by the new-fangled penny post and a lot of letter-writing. But, in general, the reach of most people's social worlds was pretty much confined to those they encountered in person. It seems that the social networking sites have broken through the constraints of time and geography that limited people's social world in Darwin's day.

One of the curious by-products of this technological revolution has been a perverse kind of competition about the number of friends you have on your personal site. Some of these claims have been, to say the least, exaggerated, with the number of registered friends running into the tens of thousands in some cases. However, even a cursory glance around this odd little electronic world quickly tells us two things. First, the distribution of the number of friends is highly skewed: most people have a pretty average number of 'friends' on their list, with only a handful having numbers above two hundred. Second, there is an issue about what really counts as a friend. Those who have very large numbers – that's to say, larger than about two hundred – invariably know little or nothing about most of the individuals on their list.

To begin at the beginning

The opening words of Dylan Thomas's *Under Milk Wood* introduce us to the small, rather dubiously named Welsh fishing community of Llareggub (for those who don't know, try reading it backwards) whose intertwined relationships wind through his drama like the ribbons on the maypole at the end of the dance. Each individual has his or her place in the social fabric of that small inward-looking world. Each has secrets that would tear that world asunder if they ever came out. In this, we are simply asserting our primate heritage – a heritage of deep social complexity involving personal relationships that, by the standards of more sensible mammals and birds, are unusually tangled and interdependent. And that primate heritage begins with the fact that monkeys and apes have much bigger brains for body size than any other group of animals.

So why do primates have such big brains? There are two general kinds of theories. The more traditional view is that they need big brains to help them to find their way about the world and solve problems in their daily search for food. The alternative view is that the complex social world in which primates live has provided the impetus for the evolution of large brains. The main version of this social intelligence theory, once known as the Machiavellian intelligence hypothesis, has the merit of identifying the thing that sets primates apart from all other animals – the complexity of their social relationships.

Primate societies seem to differ from those of other animals in two key respects. The first is the dependence on intense social bonds between individuals, which gives primate groups a highly structured appearance. Primates cannot join and leave these groups as easily as animals in the relatively unstructured herds of migrating antelope or the swarms of many insects. Other species may have groups that are highly structured in this way – elephants and prairie dogs are two obvious examples – but these animals differ from primates in a second respect. This is that primates use their knowledge about the social world in which they live to form more complex alliances with each other than do other animals.

This social intelligence hypothesis is supported by a strong correlation between the size of the group, and hence complexity of the social world, and the relative size of the neocortex – the outer surface layer of the brain that is mainly responsible for conscious thinking – in various species of nonhuman primates. This result seems to reflect a limitation on the number (and/or quality) of relationships that an animal of a given species can keep track of simultaneously. Just as a computer's ability to handle complex tasks is limited by the size of its memory and processor, so the brain's ability to manipulate information about the constantly changing social domain may be limited by the size of its neocortex.

In evolutionary terms, the correlation between group

size and neocortex size suggests that it was the need to live in larger groups that drove the evolution of large brains in primates. There are several reasons why particular species might want to live in larger groups, not least protection against predators. And it is conspicuous that the primates which both live in the largest groups and have the biggest neocortices are species such as baboons, macaques and chimpanzees, which spend most of their time on the ground and live either in relatively open habitats such as savannah woodlands or on the forest edge, where they are exposed to much higher risk from predators than most forest-dwelling species.

Dunbar's Number

This relationship between neocortex and group size in the nonhuman primates raises an obvious question. What size of group would we predict for humans, given our unusually large neocortex? Extrapolating from the relationship for monkeys and apes gives a group size of about 150 – the limit on the number of social relationships that humans can have, a figure that is now graced by the title Dunbar's Number. But is there any evidence to suggest that groups of this size actually occur in humans?

On the face of it, things do not look promising. After all, in the modern world, we live in cities and nation states that contain tens of millions of individuals. However, we have to be a little more subtle: the relationship for nonhuman primates is concerned with the number of individuals with whom an animal can maintain a coherent face-to-face relationship. It is quite obvious that those of us living in, for example, London do not have personal relationships with every one of the other ten million who live there with us. Indeed, the vast majority of these people are born, live and die without ever knowing each other's names, let alone meeting. The existence of such large groupings is certainly something we have to explain, but they are something quite different from the natural groupings we see in primates.

One place we might look for evidence of 'natural' human group sizes is among pre-industrial societies, and in particular among hunter-gatherers. Most hunter-gatherers live in complex societies that operate at a number of levels. The smallest groupings occur at temporary night camps and have between thirty and fifty individuals. These are relatively unstable, however, with individuals or families constantly joining and leaving as they move between different foraging areas or water holes. The largest grouping is normally the tribe itself, usually a linguistic grouping that defines itself rather strictly in terms of its cultural identity. Tribal groupings typically number between five hundred and 2,500 men, women and children. These two layers of traditional societies are widely recognised in anthropology. In between these two layers, however, is a third group often discussed, but seldom enumerated. Sometimes it takes the form of 'clans' that have ritual significance, such as the periodic celebration of coming-of-age ceremonies. Sometimes, the clan is based on common ownership of a hunting area or a set of water holes.

For the twenty-odd tribal societies where census data are available, these clan groups turn out to have a mean size of 153. The sizes of all but one of the village- and clan-like groupings for these societies fall between one hundred and 230, which is within the range of variation that, statistically, we would expect from the prediction of 150. In contrast, the mean sizes of overnight camps and tribal groupings all fall outside these statistical limits.

But what about more technologically developed societies? Is there anything to suggest that the figure of 150 might be a relevant social unit? The answer is yes. Once you start to look for them, groups of about this size turn up everywhere. My colleague Russell Hill and I asked a number of people to make a list of all those to whom they sent Christmas cards. On average, sixty-eight cards were sent to households that contained a total of around 150 members.

The same figure turns up in business. A rule of thumb commonly used in business organisation theory is that organisations of fewer than 150 people work fine on a person-to-person basis, but once they grow larger than this they need a formal hierarchy if they are to work efficiently. Sociologists have known since the 1950s that there is a critical threshold in the region of 150 to two hundred, with larger companies suffering a disproportionate amount of absenteeism and sickness. Famously, Mr Gore, the founder of GoreTex, one of the most successful of all medium-sized companies, insisted on creating completely separate factory units each with about 150 workers rather than just making his main factory larger when the growth of his business demanded more production – something that I suspect was the key to the success of his enterprise. By keeping his factory units below the critical size of 150, he was able to do away with hierarchies and management structures: the factory worked by personal relationships, with a sense of mutual obligation encouraging workers and managers to co-operate rather than compete.

Military planners seem to have come up with the same rule of thumb too. In most modern armies, for example, the smallest independent unit is the company, which normally consists of three fighting platoons of thirty to forty soldiers each plus the command staff and some support units, making a total of 130–150. Even the basic fighting unit of the Roman army during the Republic (the maniple, or double century) was of similar size, containing roughly 130 men.

Even academic communities may be limited in the same way. In a survey of twelve disciplines from both the sciences and the humanities, Tony Becher of the Education Department in the University of Sussex found that the number of researchers whose work an individual was likely to pay attention to was between one hundred and two hundred. Once a discipline becomes larger than this, it seems that it fragments into two or more sub-disciplines.

In traditional societies, village sizes seem to approximate this, too. Neolithic villages from the Middle East around 6000 BC typically seem to have contained 120 to 150 people, judging by the number of dwellings. And the estimated size of English villages recorded by William the Conqueror's henchmen in the *Domesday Book* in 1086 also seems to have been about 150. Similarly, during the eighteenth century the average number of people in a village in every English county except Kent was around 160. (In Kent, it was a hundred . . . I wonder what that tells us about the folk there?)

The Hutterites and the Amish, two groups of contemporary North American religious fundamentalists who live and farm communally (the one in the Dakotas, the other in Pennsylvania), have average community sizes of around 110, mainly because they split their communities once they exceed 150. What is interesting is the reason the Hutterites themselves give for splitting communities at this number. They find that when there are more than about 150 individuals, they cannot control the behaviour of the members by peer pressure alone. What keeps the community together is a sense of mutual obligation and reciprocity, and that seems to break down once community size exceeds about 150. Since their whole ethos is against having hierarchies and police forces, they prefer to split the community before they get to that point.

One way of defining Dunbar's Number is as the set of people who, if you saw them in the transit lounge during a 3 a.m. stopover at Hong Kong airport, you wouldn't feel embarrassed about going up to them and saying: 'Hi! How are you? Haven't seen you in ages!' In fact, they would probably be a bit miffed if you didn't. You wouldn't need to introduce yourself because they would know where you stood in their social world, and you would know where they stood in yours. And, if push really came to shove, they would be more likely than not to agree to lend you a fiver if you asked.

So social a brain

Is this apparent cognitive limit on the size of human groups a reflection of a memory overload problem (we can only remember 150 individuals, or only keep track of all the relationships involved in a community of 150) or is the problem a more subtle one – perhaps something to do with an information constraint on the quality of the relationships involved? Let me give just two bits of evidence that point to the second as the more likely.

One of these derives from the fact that it is extremely common in primates for there to be a relationship between a male's dominance rank and the number of females with whom he is able to mate. One prediction we can make off the back of the social brain model is that the correlation should be much poorer in those species which have a relatively larger neocortex because they can use their big computers to find ways round simple dominance-based strategies. Hence, we should find a negative correlation between neocortex volume, on the one hand, and the correlation between male rank and mating success, on the other. And this is exactly what we see in the data for monkeys and apes. Lower-ranking males in species with relatively large neocortices are able to undermine the dominance of high-ranking males and get the females to mate with them. They do this by exploiting more subtle social strategies – forming coalitions with other males to undermine the power-based ranks of dominant males, exploiting female preferences, and so on.

The second example comes from an analysis carried out by Dick Byrne of St Andrews University. He and his colleague, Andy Whiten, had put together an extensive catalogue of examples of tactical deception from the literature on primates. Tactical deception is the term used to refer to cases in which one animal exploits another to gain an objective. Species that have bigger neocortices do more tactical deception.

One of the classic examples of tactical deception is the case of the female hamadryas baboon deceiving her male. Hamadryas baboons live in harem-like family units (a male with one to five females), with ten or fifteen of these family units making up a band that lives and stays together. The males are fiercely protective of their females, and will not tolerate them getting near to other males. They do this by punishing the females if they stray too far from them, and particularly if the female allows another male to get between her and the harem male. The Swiss zoologist Hans Kummer once watched a female spend twenty minutes inching her way from where the rest of her family unit was feeding to get behind a big rock. Behind the rock there was a young male from a neighbouring unit, and once there she started to groom with him. It seemed to Kummer that, while the female was behind the rock grooming this young male, she made a very concerted effort to make sure that her head was always visible to her male above the rock as he continued feeding some metres away.

There are two possible interpretations of her behaviour. From a strictly behaviourist point of view, you might argue that she was worried about the consequences of her action, having learned that not keeping within her male's view invited trouble. A more generous cognitive interpretation is that she was thinking something like the following: 'As long as the old gaffer can see my head he will think I am just sitting here innocently behind a rock and so I can get away with whatever it is I am trying to do.' The suggestion of the latter interpretation is that she was manipulating the mental state of her male.

I suspect that what she was actually doing was not quite so sophisticated as the second interpretation (though such claims have become quite common among scientists who study animal behaviour and cognition in recent years). However, irrespective of which explanation is right, behaviour of this subtlety is far from unusual among monkeys and apes – and almost unheard of among any other non-primate species. In the study of animal (and human developmental) cognition, the phenomenon is now referred to as 'mentalising' – being able to understand the minds of other individuals rather than simply working in terms of simple descriptions of their behaviour. The belief is that whereas all other animals function like behaviourists have always supposed (they learn rules of behaviour), monkeys and apes have shifted gear just enough to be able to work in terms of understanding at least a little bit of the mind behind the behaviour.

Evidence of this kind pushes us towards the view that it is something about the *quality* of the relationships that is important, not just their absolute number. We find an upper limit on group size because this is the limit of the number of relationships that an animal can maintain at this level of complexity. It's not just a matter of remembering who is who, or how *x* relates to *y* and both relate to me, but rather how I can use my knowledge of the individuals involved to manage those relationships when I need to call on them.

Primates are above all social animals: that is their big evolutionary breakthrough. It's what has made them as successful as they have been and, by extension of course, it is what makes humans so successful – we have inherited the same social expertise. What marks primates (or at least monkeys and apes) out as different from all other species of animals is the sheer intensity of their social interactions. The difference between the rest of our primate cousins and us is simply that we have taken this trend to a whole new level.

Counting your friends in threes

Noah, it is said, counted the animals into his Ark two by two. Perhaps sensibly in view of the circumstances, he was no doubt thinking in terms of reproduction. Had he been thinking socially, he might instead have counted his animals by threes. That, at least, is the message of several recent studies suggesting that our social networks have a very distinctive structure based on multiples of three.

We all know that we can distinguish friends from acquaintances by how we feel about them. Friends are those we want to spend time with, whereas acquaintances are those whose company is more of a momentary convenience. But it seems that we make even finer judgements than this in real life. What's perhaps more intriguing is that if you look at the pattern of relationships within the group of 150 that constitutes our social world, a number of circles of intimacy can be detected. The innermost group consists of about three to five people. These seem to constitute the small nucleus of really good friends to whom you go in times of trouble – for advice, comfort, or perhaps even the loan of money or help. Above this is a slightly larger grouping that typically consists of about ten additional people. And above this is a slightly bigger circle of around thirty more.

The numbers that make up these circles of acquaintanceship seem to have no obvious pattern. But if you consider each successive circle inclusive of all the inner circles, a very clear pattern emerges: they seem to form a sequence that goes up by a factor of three (roughly five, fifteen, fifty and 150). In fact, there are at least two more layers beyond this: there is a grouping at about five hundred and another at about fifteen hundred. And the Greek philosopher Plato even managed to get the next layer out: he identified 5,300 (and I'll happily allow him the extra three hundred) as the ideal size for a democracy . . .

We are not sure what all of these successive circles correspond to in real life, or why they should increase in size by a multiple of three, but some correspond to very well-known groupings. The grouping of twelve to fifteen, for example, has long been known to social psychologists as the 'sympathy group' – all those whose death tomorrow would leave you distraught. Curiously, this is also the typical team size in most team sports, the number of members on a jury, the number of Apostles . . . and the list goes on. The fifty grouping corresponds to the typical overnight camp size among traditional hunter-gatherers like the Australian Aboriginals or the San Bushmen of southern Africa. And 1,500 is the average size of tribes among hunter-gatherer peoples (usually defined as all the people that speak the same language, or, in the case of very widespread languages, the same dialect).

It seems that each of these circles of acquaintanceship maps quite neatly onto two aspects of how we relate to our friends. One is the frequency with which we contact them – at least once a week for the inner circle of five, at least once a month for the circle of fifteen, at least once a year for the 150. But it also seems to coincide with the sense of intimacy we feel: we have the most intense relationships with the inner five, but we have a slightly cooler relationship with the ten additional people that make up the next circle of fifteen. And successively cooler still are our feelings towards the next two layers (those in the circles of fifty and 150).

So it seems as though there is a limit to the number of people we can hold at a particular level of intimacy. There are just so many boxes you can fill in your innermost circle, and if a new person comes into your life, someone has to drop down into the next level to make room for them. Interestingly, kin seem to occur more often than you would expect by chance in each of these successive levels. This isn't to say that we have to include (or even like!) all our kin, but it does seem that kin get given preference: when all else is equal, blood really is thicker than water and we are more willing to help them out.

Chapter 4

Kith and Kin

Community is what makes the world go round. In that respect, we are very much in tune with our primate heritage: sociality, often a very intense form of sociality, is the hallmark of the monkeys and apes. It was the big key to their – and our – evolutionary success. And the core of that sense of community – especially in humans – is kinship. Kinship provides a surprisingly deep and sometimes unrecognised framework for our social life, not just in traditional small-scale societies, but for us today as well.

In praise of nepotism

Around 1900, my grandfather left the family stronghold in Moray in the northeast of Scotland and headed east . . . to India, where he ended up in the small dusty town of Kanpur (then spelled Cawnpore), more or less in the middle of nowhere on the great Ganges Plain. In the end, he spent the rest of his life in and around the great northern plains at the foot of the Himalayas, and never went back to Scotland – though he retained throughout his life the links with home, including the little family cottage that his grandfather had built in Kingston at the mouth of that great salmon and whisky river, the Spey.

I have often wondered what made him get up and go, the only member of our extended northeastern family to leave Scotland (aside from when, a century earlier, his own grandfather had spent a year or so in Spain and, later, at Waterloo earning the king's shilling defending us against Napoleon). By chance, just a couple of years ago, I discovered the answer. It was very simple. His maternal cousin had gone there a few years before him, and had evidently fixed him up with a job with a local firm of stonemasons.

Well, that only pushes the question back one step further. So why did his cousin go to this obscure corner of the Raj? The answer lies in whom he worked for . . . the Elgin Cotton Mill. And who owned and ran the Elgin Cotton Mill? And, as it happens, the Muir Mill, the Cawnpore Cotton Mill, the Stewart Harness and Saddlery Factory and several other local industrial companies in Kanpur? Mostly, as the names might suggest, Scots from the northeast, who for various reasons had ended up in Kanpur in the aftermath of the Indian Mutiny and spotted an opening in the industrial market.

And here is the issue. When they needed to recruit staff, they invariably sent back home for them, back to their own communities, where they could get people they could trust and rely on. And they could rely on them precisely because of that sense of community, of belonging to the same small interdependent social network back home. It helped, of course, that old Granny's beady eye would be upon them even in far-off lands, that tongues back home would be set a-wagging by the mere breath of rumour should they step out of line. But these niceties aside, the ties of kinship and community pulled enough weight on their own to keep most people toeing the line.

It is a pattern that one sees over and over again in the history of Scots migration. When the founding Scots fathers of Princeton University in the fledgling United States sought a principal for their new educational establishment, they did not advertise as we would now, but sent for one of their own from Edinburgh to head up the new institution.

In short, nepotism played an important role in the history of Scots migration, and its benefits were enormous. It probably made the Scots the single most successful migrant group from the British Isles during the eighteenth and nineteenth centuries. The empire that was run from London was, in reality, a Scots empire, disproportionately administered, policed, missionised, taught, geologised, doctored, nursed, traded and transported by Scots. The issue was not so much that the Scots were any more desperate for a decently paid, upwardly mobile life than the English or the Welsh or the Irish, but that a strong sense of home community bound them to each other, and made working together that much more effective. That, and an education system second to none.

Despite the patent disapproval of his subsequent employers (the decidedly anti-British American Presbyterian Mission in North India), my grandfather continued to be a regular visitor to the British Club solely in order to spend time with the Scots officers of the regiments stationed in the locality. I hasten to add that he was a lifelong teetotaller, so it wasn't the drink that drew him there – just the social gathering and the opportunity to immerse himself for an evening in things Scottish.

The Scots have had a long tradition of such clubbishness. There had been mass migrations from Scotland to London in the second half of the seventeenth century that were associated with the founding of many Scots clubs and associations in the capital. The Highland Society was founded in London in the 1750s to provide support for immigrant Scots and, importantly, to ensure the preservation of Scottish culture, dress, music and language – and when they said language, they meant, of course, Gaelic. By the end of the nineteenth century, there were more than thirty Scottish societies, associations and clubs in the capital, many of them local county associations – the Argyllshire Association, the London Murray-shire Club, and so on – intended to maintain local community relationships as well as acting as mutual help societies.

Community, in a word, is the beating heart of life, and we neglect it at our peril. And one reason why, in traditional societies, communities were as effective as they were is that they consisted almost entirely of kin. As the Inuit whalers who take on whales in small open boats, *Moby Dick* fashion, don't hesitate to point out: when the chips are down and you've been thrown out of the boat into freezing Arctic waters, no one except a close relative is likely to be willing to put their life on the line to rescue you.

Thanks be to kin

In the modern world, we have lost that all-encompassing sense of kinship that pervades traditional small-scale societies. In these societies, everyone in the community is kin. This is not just because they invent kinship relationships, even for incomer strangers like the anthropologists who come to study them. It is because everyone really is kin, related to each other in a complex biological web.

Those who come into the community (with the possible exception of the lonely anthropologist) soon become embedded into that web of relatedness because they marry and have children with members of the community. What makes us kin is not so much that we are descended from some remote common ancestor, but rather that we share a common interest in the future generations. We refer to in-laws as relatives for the very good reason that we and they share a common genetic interest in the offspring who will, in due course, become the parents of the next generation.

The importance of kinship is well illustrated by one of the iconic events in American folklore. In May 1846 at the height of the 'taming of the Wild West' and gold fever, the intrepid colonists of the Donner Party set out from Little Sandy River in Wyoming on the last stage of a long trek to California and a new life, a journey that had begun in Springfield, Illinois, more than a month before. Several untoward events – disorganisation at the start, some ill-advised routing, and attacks by Indians along the way – conspired to delay the party, which at its height numbered eighty-seven men, women and children. As a result, they reached the Sierra Nevada mountains, the jagged line of snow-covered peaks that barred their way west, much later than they had intended, just as winter began to close in.

Though they struggled on, they ended up trapped in the mountains by snowstorms at an entirely anonymous spot now known as Donner Pass. Here, they tried to sit out the winter. But since they had expected to be through the mountains well before winter set in, they had come unprepared. Their food gave out, and some even gave in to cannibalism. By the time a series of rescue parties arrived from California in February and March the following year, forty-one of the eighty-seven pioneers had died. What makes these bald statistics interesting is who died and who survived. Disproportionately more people who travelled alone died, while the chances of surviving were much higher among those who had travelled as families. Frail grannies travelling with their families made it, but not the strapping young men travelling alone. It paid to be travelling with kith and kin.

A second example is provided by another of the iconic events in American folklore. When the *Mayflower* colonists set foot on the American mainland in 1620, they were ill prepared to face the harsh New England winter. They suffered from severe malnutrition, disease and lack of resources, and no fewer than fifty-three of the 103 colonists died in that first winter. But for the intervention and generosity of the local Indians, the colony would have died out completely. Again, mortality was highest among those who came alone, and lowest among those who came as families.

The issue is not so much that families rush around and help each other, though that is certainly true, but rather that there seems to be something enhancing about being with kin. Being surrounded by family somehow makes you more resilient than when you are simply with friends – however much you argue with them. This much is clear from two studies of childhood sickness and mortality, one in the city of Newcastle-upon-Tyne during the 1950s and the other on the Caribbean island of Dominica during the 1980s. In both cases, the amount of childhood illness and mortality experienced by a family was directly correlated with the size of its kinship network. Very young children in big families got sick less often, and were less likely to die. Again, this is not just because there are more people to rush around and do things in large families. Rather, it has something to do with just being in the centre of a web of interconnected relationships. Somehow, it makes you feel more secure and content, and better able to face the vagaries the world conspires to throw at you.

And your name is . . .?

Just how potent the sense of kinship can be is nicely shown by how influential personal names can be. Until about a century ago, the old Gaelic naming tradition still applied widely in Scotland. By these rules, the first son was named after his paternal grandfather, the second after his father, the third after a father's brother, with the equivalent rules on the maternal side applying to daughters. I actually owe my first name to a revolt on the part of my mother who flatly refused to have yet another George in the family – otherwise, if my father had had his way, I would have been the fifth George Dunbar in a row, starting with my great-great-grandfather who had been born in 1790.

But why should naming have followed these kinds of rules?

One obvious answer is that bearing the same name identifies family membership. This much is self-evident from the way we use surnames, although some surnames are clearly considerably better for this than others. While

Bakers and Smiths must sadly conclude that they are unlikely to be related to strangers bearing the same name, Gaelic family names do provide clear indications of common ancestry, partly because of their many variants. Many name lineages are of quite modest size, and many had quite localised origins. The seaport up the road from Edinburgh notwithstanding (whose castle was in fact once the seat of the family's medieval powerbase), Dunbar has been an almost exclusively Moray name for several centuries and rare elsewhere.

But it seems that first names can imply something about relatedness too. The traditional habit of naming a child for someone else seems to create a bond that invites interest and the possibility of lifelong investment by the person after whom the child is named. Traditionally, German children had one Christian name for every godparent that its parents had made the effort to ask – and godparents were expected to help further the child's interests in society once it reached adulthood, not just to worry about its attending Sunday school. Analysis of parish registers from the Krummhörn area in northwest Germany by Eckart Voland, a historical demographer at the University of Giessen, showed that children who survived the first year of life typically had more Christian names than those who did not: since names were conferred when the child was baptised on its eighth day of life, this suggests that parents already knew who would survive and who would not, and hence for which children it was worth making the effort of soliciting godparents.

This sense of implied kinship even seems to persist today. This was put to direct test in a recent study carried out by evolutionary psychologists from Canada's

McMaster University. They used the US census to select a set of common and rare English surnames and first names, and then emailed nearly three thousand Hotmail accounts with different combinations of these names asking for help with a project on local sports team mascots, ostensibly from someone with the same or different combination of names. The test was whether the recipient took the trouble of replying. Just two per cent of recipients replied when they shared neither first name nor surname, but twelve per cent did so when they shared both. Shared surnames (which resulted in six per cent of recipients replying) did better than shared first names (four per cent). But when the names were rare in the population at large, the reply rates soared to twenty-seven per cent when sender and recipient shared both names, and thirteen per cent when they shared just their surname. As many as a third of those replying when rare names were shared commented at length on the coincidence, often asking about family origins.

I recognise exactly these response patterns in my own behaviour. Finding someone with the surname Dunbar invariably arouses my immediate interest. But I am only mildly excited when I come across a McDonald – among the commonest of all Scottish surnames – even though it has been a middle name in my particular lineage for several generations, thanks to a McDonald great-grandmother.

Evolutionary biologists have long understood the significance of kinship (shared descent from a common ancestor) in animal and human biology. The essence of this is summed up in what has become known as 'Hamilton's Rule', one of the cornerstones of modern evolutionary biology, named after the late W. D. Hamilton who discovered it while still a humble PhD student in the 1960s. Hamilton pointed out that two individuals have a genetic interest in each other that is proportional to the likelihood of their sharing a given gene by descent from a common ancestor, and hence that, when all else is equal, they should be more likely to behave altruistically towards each other than individuals who are less closely related. Blood, as the old saying goes, is thicker than water. It is a finding that has been widely demonstrated by observation and experiment in organisms ranging from tadpoles to humans.

Our naming patterns seem to capitalise on this. In fact, the biological intuition of relatedness seems to be so strong that, in the absence of anything else, shared names can be used to trigger sentiments of kinship even where none actually exists.

Names are not the only way we identify family connections. Dialects are another. Dialects are actually rather odd things. Language, as we may reasonably assume, evolved to enable us to communicate with each other so as to get the communal jobs done better. Yet languages have an extraordinary capacity to fractionate into mutually unintelligible dialects at an astonishing rate – on a scale of generations rather than millennia. Not to put too fine a word on it, generations are parts of a population separated by language. But why on earth would something designed to make communication possible have the inherent property of preventing mutual comprehension?

The answer to this evolutionary conundrum is that dialects are a very reliable marker for your place of birth. Even as recently as the 1970s, it was possible to place a native English speaker to within thirty miles of his or her birthplace. In effect, because it is learned young and cannot easily be learned later in life, dialect provides a useful cue of which community you were born into, and so whom you are likely to be related to. It is one of many kinds of social badge that we use to identify membership of a local community, and thus who can be relied on and to whom one should owe obligations. In a study carried out by Jamie Gilday (then a student in our lab), people phoned at random were more likely to agree to help complete a task over the phone if they had the same local accent as the caller (raised in a Lanark village) than those whose dialects were markedly different (city Glasgow or northern English). In another study, my one-time graduate student Daniel Nettle showed that so long as dialects changed fast enough, they could prevent freeloaders who exploited social obligations from taking over a population.

Chapter 5

The Ancestors that Still Haunt Us

It is a truism to say that your past is in your genes. But for all its humdrum dullness, the fact is that modern genetics has uncovered some fascinating insights into our recent past that we could never have gleaned from the history books. The DNA of our chromosomes is literally the history of our individual ancestries. Although we receive half our genes from each of our parents, some genes are transmitted only through one sex. The Y chromosome is passed on only from father to son, and identifies uninterrupted male lineages. In contrast, mitochondrial genes are inherited only from your mother. The mitochondria are the tiny powerhouses that fuel a cell's activities. In the very remote past, they were free-living viruses that found a cosy home inside the cells of 'proper' animals; there they set up home within the cytoplasm that surrounds the nucleus where the chromosomes are housed. As a result, they are passed on only in the egg, and so always come from your mother. They allow us to track maternal lineages.

Descended from the Khan?

If your family name happens to be Khan, there's a fair probability that you are descended from the greatest of all Khans, the warrior king Genghis Khan whose Mongol armies swept through central Asia as far as Tashkent and northern Pakistan in the first decades of the thirteenth century. But even if your surname isn't Khan, no need to be disappointed: modern genetics has shown that there's still a fair chance that you are a descendant of the Khan. A recent survey of Y-chromosome genes revealed that an astonishing 0.5 per cent of all the men currently alive today have inherited their Y chromosome from the great Mongol warrior or his brothers. And if your ancestors are from the central Asian heartland of the old Mongol empire, that chance rises to one in twelve (8.5 per cent) of all men.

These extraordinary findings come from a study of the DNA of over two thousand men sampled from right across central Asia, from Japan to the Black Sea. While the Y chromosomes of most of the men in the sample showed the usual wide range of DNA types (known in the trade as haplotypes), nearly two hundred shared a set of very similar (sometimes identical) genetic signatures. This set of about eighteen haplotypes formed a very distinct cluster that set them apart from the other sixty or so haplotypes in the sample.

The research team was intrigued by two facts about this unusual cluster of haplotypes. First, they formed a particularly dense concentration in the region of modern Mongolia; second, there were pockets of them all over central Asia. In contrast, all the other haplotypes were very localised to particular hotspots.

Evolutionary theory offers us three possible explanations as to why a genetic lineage might be both as common and as geographically widespread as this. One is that it might arise simply by chance, and, being of no particular advantage or disadvantage to those who happened to inherit it, it spread gradually by a process known as genetic drift. A second is that the genes in question have been particularly advantageous and so have been subject to intense selection. The third is a form of sexual selection, whereby males who possessed these haplotypes were unusually successful in reproducing themselves.

A few quick calculations are enough to suggest that the first is unlikely: even by the most conservative estimates, the chances of such a distribution arising by chance are considerably less than 100 million to one. The second is not a lot more plausible: the Y chromosome is tiny and contains almost no genes other than those required to turn the foetus into a male (more on this later). That leaves us with the third possibility. And here history comes to the rescue. A glance through its pages quickly identifies one event that might just fit the bill: Genghis Khan's empire.

Two pieces of the jigsaw make this explanation plausible. One is the fact that all the haplotypes in the cluster come exclusively from the areas that came under the great Khan's rule. The unusual haplotypes are completely absent elsewhere in Asia that remained outwith the Mongol empire. The second is the time of origin of this cluster of haplotypes. Many of our genes have no function (i.e. don't code for the proteins involved in actually building the body) and so they only change over time as a result of random mutations. This has allowed biologists to use them as a kind of molecular clock: count the number of neutral or 'junk' genes by which two individuals differ, divide by the rate at which genes mutate, and – hey presto! – we have a very reasonable estimate of how long ago they last shared a common ancestor. When the researchers did this for the eighteen or so haplotypes in their unusual cluster, they arrived at a figure of 860 years ago. Genghis Khan was born around AD 1160, just 840 years ago. This is close enough to be suspicious, as Sherlock Holmes might have said. More interestingly, perhaps, it suggests that the original mutation that produced the haplotypes in question derived not from the Khan himself, but a generation earlier in his lineage – from the Khan's father, Yesugei.

When Yesugei's young son Temüjin united the factious Mongol tribes in 1206 and earned himself the title Genghis Khan – 'Khan' meaning ruler or emperor – he brought under his control a formidable fighting force. In a series of lightning strikes, he conquered the two northern Chinese empires and then struck westwards through modern-day Kazakhstan as far as the Black Sea to create the biggest empire in all history. Although invariably heavily outnumbered, his battle-hardened troops demolished everything the opposition placed in his way.

And then – to use his own words – having vanquished your enemies, 'the greatest happiness is to chase them before you, to rob them of their wealth, to see those dear to them bathed in tears, to clasp to your bosom their wives and daughters'. It seems, from modern genetics, that the Khan and his brothers were as good as their word.

Pity the poor Basques

As that great assertion of Scots independence, the Declaration of Arbroath, succinctly put it in 1320: the Scots 'journeyed from Greater Scythia . . . to their home in the west where they still live today'. And just who *were* the Scythians, then? Well, actually, a very successful group of pastoralists who first appeared in the western edges of Mongolia around 3000 BC and gradually made their way westwards, spending time on the way in what is now Uzbekistan near the Aral Sea, then in the Caucasus region of Georgia, finally entering eastern Europe through the Ukraine.

Are the Scots *really* the descendants of the Scythians? Well, actually probably not – it was more of a political move to persuade the Pope, to whom the Declaration was addressed, that the Scots could not possibly ever have been English, and therefore should not have to be the vassals of the English king Edward II. But, as far-fetched as this specific claim might have been, it seems that the authors of the Declaration were not quite as far off the mark as all that – though how on earth they could have known that is another question. Most of us Europeans are in fact the descendants of the great Indo-European expansion that began around 3000 BC somewhere in the steppes of southern Russia. The Scythians were, to be fair, a rather late element in that story, and probably never did get much further than the Ukraine. But the wonders of modern genetics tell us that the great Indo-European invasion served to displace (or worse) most of the previous inhabitants of Europe over the next couple of thousand years. Today, all but a handful of the myriad languages spoken in Europe are descended from the language spoken by those early Indo-European immigrants.

It seems that only the Basques survived this human tsunami with anything like their national identity – or their genes – intact. Protected by their Pyrenean mountain fastness, the ancestors of the Basques must have watched the tidal surges of successive invasions and conquests that lapped the foothills of their mountain home with, shall we say, concern. But, by a quirk of geography, they survived relatively unscathed, aloof from the turmoil that changed the face of Europe.

At least, this is the conclusion that we are drawn to by the converging evidence from both linguistics and genetics. Linguists have known for a long time that the Basque language is an oddity. It is completely unrelated to – and quite unlike – any of the other languages of Europe, which, with the exception of a handful of exceptions, are all part of the great family of Indo-European languages. (Among the best known of these exceptions are Finnish and Hungarian, both of which derive from invasions by Mongolian peoples, the latter most famously associated with Attila the Hun and his chums.) Indo-European is a language family that spans the Gaelic tongues of the far west, almost all the other languages of modern Europe, the Farsi and Pushtu languages of modern Iran and Afghanistan, Sanskrit and Urdu and their many descendants in northern India, and reaches its easternmost extension with Bengali in Bangladesh. The closeness of these languages is reflected in the similarity of many of their everyday words. The Indian Sanskrit word *bhrater* is recognisably but a shade away from the Gaelic *bràthair* and the English *brother*, and manifestly a world away from, for example, the East African Swahili equivalent *kaka*. Unlike Swahili, Sanskrit, Gaelic and English share a recent common ancestry in the great Indo-European expansion.

Basque is the one European exception. It shares almost nothing in common with any of the Indo-European languages, as is evident from the Basque word for 'brother' – *anaia*. As a language, it seems to be a complete outlier, although some linguists have claimed that its nearest language relatives are some small pockets of relict Caucasian languages scattered across the southern steppes of Russia, which in turn form part of the Dene-Caucasian family of languages. What makes this family interesting is the fact that the Dene half consists of the Na-Dene American Indian languages, spoken in a wedge straddling the present Canadian–US border inland from the Pacific coast about as far east as the Great Lakes. We would have to go back a very long way to find the common link between the Indo-European and Dene-Caucasian language families.

Genetics has given us a new window on this intriguing story. Once again, the Basques seem to sit aside from the rest of Europe, an isolate with few genetic links to anyone else in Europe, although they share some gene complexes with the early Celts (even though these were part of the early Indo-European expansion). Let me illustrate this with just one example. The incidence of the rhesus negative gene in modern Indo-Europeans is just about two per cent, and four to eight per cent in African Americans. But it is nearly thirty-five per cent in the Basques, and around fifteen per cent in Caucasians (that is, the people of the Caucasus with whom the Basques may share linguistic origins). So the Basques might be the last misty flicker of the original inhabitants of Europe just before our own Indo-European ancestors turned up. Some have even suggested that it was the Basques' ancestors that were responsible for the truly astonishing inflorescence of cave paintings in northern

Spain and southern France between twelve thousand and thirty thousand years ago.

So in these days of angst about homelands and migrants, we might spare a thought for the Basques, Europe's original inhabitants. Which leads to an interesting thought. If the Basques really are the remnants of the original inhabitants of Europe, might they legitimately lay claim once more to the continent? What should we do if they asked, politely no doubt, whether the rest of us would mind getting back to southern Russia where we came from?

My dad was a Phoenician

Wholesale migrations probably tend to result in the extinction or displacement of the luckless folk who find themselves in the way of a migrant wave. Something along these lines occurred in Europe when the Indo-Europeans turned up from further east and pretty much forced the original inhabitants of Europe westwards, where it is thought that, like the Basques, they might still survive in isolated, barely recognisable pockets. This is suggested by the fact that the Indo-European genetic signal declines in concentration from east to west in modern Europe. In more recent historical times, of course, much the same happened in North America and Australia, where the original inhabitants have been reduced to small, socially and economically isolated communities whose long-term prospects as a distinct ethnic community are probably bleak.

Trade and military conquest, however, tend to produce different signatures. They rarely result in wholesale extinction of local communities, but traders and invaders often leave traces behind them. Since most traders and soldiers are male, it is inevitably the case that these traces are most obvious in the Y chromosome.

Sometimes, the people themselves are actually aware of their heritage. In northern Pakistan, for example, the Burusho, Kalash and Pathans all claim to be the descendants of Greek soldiers from Alexander the Great's all-conquering army in 326 BC. Pakistan represents the furthest east that Alexander managed to get in his whirlwind conquests. Considering that he and his army weren't around for all that long (mainly thanks to young Alex's untimely death at the tender age of thirty-two), it is remarkable that they left anything more than their name deeply scarred into the invariably ravished and pillaged populations they conquered. Nonetheless, a recent analysis of the genes of around a thousand Pathan men turned up a handful of individuals who have particular genes that otherwise occur in significant numbers only in modern-day Greece and Macedonia. The traces are weak, but they are there. The folk legends seem to be true.

Trade, as opposed to conquest, was what motivated the Phoenicians during much the same period. For the better part of a thousand years between around 1500 and 330 BC, Phoenician galleys traded widely throughout the Mediterranean from their homeland in modern-day Lebanon and western Syria. But by the time the Romans turned up on their patch in the closing centuries of the pre-Christian era, they had disappeared. They left relatively little trace of their existence other than in contemporary histories (including the Bible, of course) and in the fact that they produced one of the earliest alphabets. The Canaanite–Phoenician alphabet is the direct ancestor of many of the modern alphabets. The Phoenicians never aimed at conquest, but instead simply established trading colonies all over the Mediterranean – there have even been suggestions that they made it as far as the British Isles.

Recently, a rather sophisticated analysis of male Y-chromosome genes sampled all over the Mediterranean basin has managed to uncover what seem to be some specific Phoenician genetic lineages. Those parts of the chromosome that do not seem to have a direct function (i.e. don't code for the proteins involved in actually building the body) tend to have higher mutation rates than the bits that really matter, and over time these tend to come to characterise certain male lineages in particular localities. By focusing on those locations that were known to be Phoenician trading strongholds (the list included Crete, Malta, Sardinia, western Sicily, southern Spain and coastal Tunisia) and comparing them both with nearby sites with no historical record of a Phoenician presence and with sites that the Greeks colonised later, the study was able to show that a handful of distinctive Y-chromosome types were probably of Phoenician origin. In case you happen to have them, they go under the rather uninspiring names of J2, PCS1+, PCS2+ and PCS3+. If you have one of these, be in no doubt: your Dad was a Phoenician.

Slaves to the past

Slavery has been much in the news recently, not least thanks to the fact that 2007 was the two hundredth anniversary of the slave trade's abolition in Britain. Nonetheless, amid all the fuss, we risk obscuring the fact that slavery has a very ancient history, as well as a recent one. Britons also forget, perhaps, that their islands have, as much as anywhere else, been subject to the forcible removal of their inhabitants for a life of slavery elsewhere for as long as history has anything to tell us. While the inhabitants of Scotland were no doubt spared much of this, not a few of their fellow Celts from England found their way involuntarily to Rome during the long Roman occupation of Britain. It is thought that between a quarter and a third of all the people who lived in Italy at the height of the Roman Empire were slaves. Rome's economy was entirely dependent on slave labour, and they came from all over the known world.

Things did not improve all that much for the beleaguered inhabitants of these islands after the departure of the Romans. In the two centuries or so after the somewhat precipitate departure of the legions in AD 410, their replacement by a motley collection of Angles, Saxons, Frisians and Jutes from across the North Sea merely added to the woes of the Romano-British and Celtic inhabitants who had been left to fend for themselves.

Studies of the genetic make-up of the southern English of today reveals that Celtic genes become increasingly rare, and continental Anglo-Saxon genes progressively more common, as you go from the Welsh Marches to East Anglia. However, while as many as fifty per cent of the Y chromosomes carried by the inhabitants of the south-east are of continental Anglo-Saxon origin, this seems not to be true of female genes. Computer simulations carried out by Mark Jobling and his colleagues at University College London suggest that a relatively small number of Anglo-Saxon men had more than their fair share of the local Celtic women, very much to the exclusion of the local Celtic men. History offers some hints of what might have happened: the name 'Welsh', for example, derives from the Anglo-Saxon *wealasc*, which has been variously translated as 'foreigner' and 'slave' (which, to the incoming Anglo-Saxons, probably meant much the same thing). Indeed, the *wealasc* didn't even have the same rights under the law as Anglo-Saxons, and it took the better part of five hundred years before this unexpectedly ancient form of apartheid was lost in both society and the law.

Although the Scots and Irish didn't have quite as much trouble from the Romans and the Anglo-Saxons, in fact their relative immunity from outside interference didn't last all that much longer. The give-away lay hidden in the genes of the Icelanders for the better part of ten centuries until modern geneticists turned their eyes on this historically isolated community. Rather to their surprise, they discovered that, while Icelandic Y chromosomes come from fairly conventional Norwegian and other Scandinavian stock, an astonishing fifty per cent of Icelandic women's genes have Celtic origins. And guess where they came from? Yes, Scotland and Ireland – a convenient spot to stop off and pick up some women on one's way to a new life in Iceland. Especially if one's own Scandinavian women weren't so keen on the rather grim sea voyage on offer and the prospect of a hard life on a volcanic outcrop.

All this raises some interesting questions about history and how we see it. Should the Scots and Irish, for example, be asking for their women back? The recent financial crises that hit Iceland notwithstanding, I rather fancy the last place the Icelandic women would really want to come back to is the grim British Isles. So perhaps they should be asking for restitution and compensation, instead – but from whom? Their genetic and social stake in modern Iceland, thirty generations later, is much too great for it to make any sense to be seeking restitution. In any case, what does it actually mean to say that the women of Iceland are half-Celt? How should their Norse half feel? Presumably, they would prefer to stay.

And what about the descendants of the British slaves hauled off to staff the villas of Roman dignitaries in far-off Italy a thousand years earlier? This far down the line it hardly matters, even if most of their descendants have probably remained in the lower strata of Italian society ever since. They are all Italians now. History and one's ancestry is fascinating to explore, but not a recipe for hand-wringing and angst. It's the future, and one's place in it, that counts.

Chapter 6

Bonds that Bind

We are, as a species, rather an uptight lot: we don't like being touched. Well, perhaps I will rephrase that. We don't like being touched by all and sundry. That's no doubt because touch is the most intimate of all the senses. A touch is worth a thousand words. We get so much more information about someone's real meaning and intentions from the way they touch us than from anything they could ever possibly say in words. Words are fickle, open to abuse, double meaning and downright deceit – all too often, they say what we don't really mean. But the intimacy of touch catapults communication between us into another dimension, a world of feeling and emotion that words can never penetrate.

Touch me tender

We engage in many forms of intimate touch – cuddling, stroking, petting, patting. These share much in common with the grooming that takes up so much time among monkeys and apes. Contrary to popular imagination, monkey grooming is not about removing fleas. It is not even just about removing the bits of debris and vegetation that clog up the fur during a day's foraging, though it certainly does this. Rather, it is about the intimacy of massage. The physical stimulation of the skin triggers the release of endorphins in the brain. Endorphins are a family of endogenous opioids that are chemically closely related to morphine and opium. They are the brain's own painkillers – part of the pain-control mechanism that cuts in when pain is low-level but chronic. Harsh, sharp pain is neutralised by the two neural pain circuits, the fast and slow circuits. In contrast, low-level pain associated with generalised stresses – such as those that come from jogging and routine physical exercise, or from mental stress – is dealt with by the endorphin system. It's what creates that sense of wellbeing and relaxed contentedness after your morning jog or that hot shower on the back of the neck. You may have noticed that if, as a habitual jogger, you can't for some reason fit in your morning jog, the day isn't quite the same, and your friends probably find you rather more tetchy than usual. It's because you've not had your morning fix, and are suffering a very mild form of cold turkey.

As with all monkeys and apes, touch is still very important to us. We have this intense desire to stroke and touch those to whom we feel close. We can't help it. It's the first thing we want to do in any kind of close relationship. There is something intensely intimate about touch – even just holding hands, or placing an arm round someone. A touch that has no emotional commitment behind it is just plain obvious. It's not for nothing that we refer to someone as being as 'cold as a fish'. It doesn't matter what the person may be saying, the lack of warmth or caring intimacy is transparent.

Touch plays – and has surely played – a much more

important role in our social lives than we ever give it credit for. One reason for this is probably that it is perceived at a deep emotional level, rather than being something we actively think consciously about in words. We do not know how to say it, but we know exactly how to interpret the meaning of a touch. It is visceral, a gut instinct, something very ancient and primitive that is buried deep down within our psyche. It is not especially well connected to the evolutionarily more recent language centres in the brain's left side. It is emotional and right-brain.

For this reason, perhaps, we tend to underestimate the importance of touch in our lives. To be fair, there is perhaps a good reason for this. Being so tightly tied in to the emotional brain, touch seems able to arouse us very easily, and it can just as quickly spill over into sex. There you are, not especially interested; and then a few caresses or a kiss that lasts just a moment too long, and suddenly the whole system flips without warning from one state of mind to the other. How often have you said: I didn't mean to, but . . .?

Perhaps that is why we are so reluctant to be in close contact with strangers, or even those we know we don't have an especially intimate relationship with. Physical contact can too easily spill over into areas of our psyche where, in cooler, more considered moments, we might not want to go. So rather than risk that sudden, uncontrolled, emotional flip, we back off and distance ourselves.

In whom we trust . . .

Every day, you drive to work, and you trust that other motorists will abide by the rules, stay on their side of the road and try not to run you down. It may seem obvious, but we take the role that trust plays in regulating our lives for granted. In fact, our entire social world depends on it. It's famously true that the diamond market in Amsterdam – the largest in the world – works entirely on what was once referred to a 'gentleman's bond'. Millions of pounds' worth of diamonds are traded solely on the strength of a handshake as to quality and payment. To be fair, there would probably be an issue of hooded gentlemen and broken legs if anyone did try to pull a fast one. But the real core to it is personal trust within a very small, closed community of fewer than a couple of dozen people. They will only trade with each other, and if you are not one of them, forget it . . . you won't even get a peek at the stuff worth buying.

Trust permeates every aspect of our daily lives. Not to put too fine a point on it, it reaches the parts that only certain beers are said to reach. There has always been an implicit assumption that trust is based on some kind of reciprocity – you scratch my back and I'll scratch yours later. Now it seems that trust has a chemical basis. The chemical in question is an obscure little thing called oxytocin. A group of economists at Zurich University in Switzerland have recently shown that a squirt of oxytocin from a nasal spray can make you more willing to share a reward with another player.

In these experiments, one contestant (the investor) was given a sum of money and then invited to share some, all or none of it with a second contestant (the trustee). Whatever the investor gave to the trustee was doubled, and the trustee then invited to share some, all or none of the enlarged pot with the original investor. The investor's risk, of course, is that the trustee simply pockets the lot. But if the investor could trust the trustee, they would both do best by the investor putting the whole sum into the initial pot and the trustee offering half of the augmented pot back. Most investors, however, hedge their bets and offer something, but not the whole lot.

But investors given a single shot of oxytocin before they made their offer shared seventeen per cent more of their initial pot with the trustee than those sprayed with an inert chemical (a placebo). What makes it clear that this is about trust is the fact that when the experiment was re-run with the trustee's decision being made at random by a computer (but with the same probability of defection – pocketing the money – as that shown by the trustees in the previous experiment), there was no difference between the oxytocin and placebo conditions in the investors' willingness to share. In other words, it was not simply risk that the investors were betting on, but their understanding of human behaviour.

What makes this experiment particularly interesting is that oxytocin turns up in other important social contexts. It is released in copious quantities during and after sex, generating that sense of deep attachment that seems to permeate every corner of our bodies in the aftermath. Comparisons of monogamous and polygamous vole species suggests that the pairbonding that underpins monogamy in these species is also based on an especially high sensitivity to oxytocin. It also facilitates nest-building and pup retrieval in rats, and mother-offspring bonding in sheep.

This does not, of course, mean that our lives are regulated entirely by chemicals. Rather, the point is that these chemicals create a neural environment that is sensitive to certain kinds of cues when these are encountered. There are many other familiar examples. We have known for over half a century, for example, that the 'flight/fight' response is underpinned in much the same way by the hormone epinephrine (aka adrenaline): release of the hormone prepares the body for action, but which action (flight or fight) depends on how the individual perceives the circumstances.

By the same token, in the Zurich experiment, some of the investors who were given oxytocin were a great deal less generous than some of those in the control group. That probably reflects a combination of two supplementary effects. One is likely to be individual differences in sensitivity to oxytocin's effects: women, for example, are more sensitive to it than men, and there will be further variation within each sex. The other is likely to be investors' sensitivities to the cues of honesty given off by the trustees once they have been primed by the hormone to pay attention to them.

Laughter, the best medicine

I once took part in a management consultancy event in London that drew together a collection of around sixty individuals from all walks of business and government life. After the inevitable croissant-and-coffee breakfast, we were herded into a side room and asked to take a seat. The chairs had been set out in circles so that everyone faced the centre of the room. We sat for maybe five minutes or so in silence, with everyone becoming increasingly edgy and puzzled about what was going on.

Eventually, one by one – but always after several minutes of intervening silence – the organisers stood up and said something to the effect of 'I believe in . . . [something or other]'. This simply served to create even more edginess among the assembled throng, and nowhere more so than among a pair of rather out-of-place, primly besuited elderly gentlemen who were obviously on a skive from one of the government ministries just round the corner in Whitehall. They were clearly beginning to wonder what on earth they had let themselves in for when they could have been more usefully running the country . . .

Gradually, one or two of the audience began to join in with rather hesitant statements about their beliefs. Then someone stood up and said: 'I believe that we are all wondering what on earth is going on.' The assembled company broke into uproarious laughter. From that moment on, the atmosphere was completely different. The ice had been cracked. Suddenly, we were instantly transformed from a group of strangers into a band of brothers (well, and sisters too, of course).

Laughter, and especially communal laughter, seems to have an extraordinary capacity to create a sense of bondedness. It is not just a matter of releasing tension. You get the same effect if you go to a theatre to see a comedian. After an hour or so spent with tears streaming down your face, you emerge feeling on a high. You are relaxed, at peace with the world, full of bonhomie. Without a moment's hesitation, you turn to a complete stranger and strike up an animated conversation. In those few minutes of passing conversation, you will probably have volunteered several snippets of personal details about yourself – something you would never have considered doing an hour or so before as you waited for the show to start.

You will even become more generous towards strangers. When Mark van Vugt and his colleagues at the University of Kent asked subjects to share a sum of money they had been given with a partner, they were much more generous to an existing friend than to someone they had never met before. But if they watched a comedy video *and laughed* together, they were as generous to strangers as to a friend. In some mysterious way, laughter turns strangers into friends.

In fact, it turns out to be anything but mysterious. Laughter – and I mean deep belly laughter, not the polite titters that tinkled among T. S. Eliot's teacups – is an extremely effective releaser of endorphins, probably because the physical effort of laughter's heaving chest is quite hard work for the muscles. We have demonstrated this using pain threshold as an assay of endorphin release. We tested subjects' pain thresholds before and after watching either a boring tourist video or a comedy video in small groups. Since endorphins are part of the body's pain-control system, pain thresholds should be much higher after laughing if laughter triggers the release of endorphins in the brain. And, so it was: those who laughed a lot while watching a comedy video had an elevated pain threshold afterwards, whereas those who watched a boring video showed no change.

I suspect that laughter is a very ancient trait. It is a behaviour we share with chimpanzees, though, as the psychologist Robert Provine has observed, the form is slightly different. In chimpanzees, it has a simple *hah-uh-hah-uh-hah* series of alternating exhalations and inhalations, whereas ours is a much more vigorous series of repeated exhalations without drawing breath: *ha-ha-ha-ha*. There are two other differences. We laugh socially, whereas chimpanzees typically laugh alone – they laugh in anticipation of, or during, a social situation, especially during play, but not together at the same time as we do. The other is that we use language (in the form of jokes) to trigger laughter. How boring is a conversation that isn't peppered with one-liners?

This last was clearly a late development that appeared only after the evolution of language. But the first – the social nature of laughter – is almost certainly much more ancient, perhaps something that evolved as much as a million or so years ago in *Homo erectus*, the precursors of the first true humans. It was most likely a form of chorusing, a kind of communal singing without the words. Its function, I think, was to generate the same kind of surge of endorphins as that produced during grooming. My guess is that this kind of social laughter came on stream, built up out of more conventional chimpanzee-like laughing, to supplement grooming as a bonding mechanism once these early ancestors of ours hit the upper limits on the time they could afford for social grooming.

Still, laughter is not the only way we produce these endorphin surges.

If music be the food of love . . .

You hear the distant strains of that old familiar song and there's that moment of recognition, that tingling mixture of half-remembered emotions. For me, it might be the strains of a Buddy Holly song, or a snatch of one of Bach's Brandenburg Concertos, or the skirl of massed bagpipes. But why is it that music moves us so?

Perhaps surprisingly, music has remained until very recently one of the Cinderella areas of modern science, something too trivial for real scientists to dirty their hands with – evolutionary cheesecake, as the linguist Steven Pinker put it. And yet, as evolutionary biologists will never tire of pointing out, something that a species is prepared to devote so much time – and money! – to cannot be a trivial by-product. Whenever animals invest that much time and effort in something, it's usually because it is of fundamental biological importance.

One suggestion – made originally by Darwin himself – is that music is a form of sexual advertising, rather in the way song functions in birds. Why else, you might ask, should sheer inventiveness of composition or musical skill play such a huge role in our appreciation of music? The fact that you can get your fingers or tongue around a complicated tune obviously demonstrates the quality of your genes to a prospective mate. It seems very plausible.

Sexual selection, as Darwin pointed out nearly 140 years ago in one of his other great books, *Sexual Selection and the Descent of Man*, is an extraordinarily powerful force in evolution, well capable of picking up the most trivial traits and exaggerating them to the point where they actually become detrimental to those that possess them – at least in terms of daily survival. The peacock's tail weighs him down when he flies, and so exposes him to a greatly increased risk of being caught by a predator. The payoff comes through greater success in the mating stakes. What the males are saying, in effect, is: *Watch me – I'm so good I can handicap myself with all this and still beat the predators!* Males with flashy tails and more eyespots do so much better in the business of attracting lady peafowl. It is but one of many well studied examples from the animal world.

And there is a certain amount of evidence to support the proposal that music functions in this way for us, not least the self-evident sexual attractiveness of pop stars. The evolutionary psychologist Geoffrey Miller found that jazz musicians, pop musicians and classical composers are all most productive during the sexually active phase of their lives. And it is no accident, surely, that Vivaldi's efforts were exercised so strenuously on behalf of the young ladies of Venice's Ospedale della Pietà orphanage – many of whom found rich husbands thanks to the skills exhibited in the concerts they performed under his direction.

To test Miller's hypothesis more precisely, one of my students, Kostas Kaskatis, looked at the productivity of nineteenth-century classical European composers – everyone from Beethoven to Mahler – and 1960s-vintage rock stars. He found that the number of new works they composed dropped off dramatically after they had married, but then rose again as soon as they had separated or divorced and were once more on the prowl for a new mate. And once they had found someone new . . . yes, it dropped off again.

Well, it may be so. But another possibility is that music had its origins in social bonding. There is something raw and primitive about music's ability to stir the emotions. Every parade-ground martinet knows that songs are the best way to create a sense of camaraderie in a group of raw recruits.

Recent brain-scan studies indicate that music seems to stimulate deeply primitive centres at the front end of the right hemisphere of the brain. In rather crude terms, the left half of your brain is more active in conscious processes – hence the fact that it is particularly strongly implicated in language – whereas your right half is more active in those unconscious, more primitively emotional aspects of behaviour.

Another recent finding is that music triggers the release of endorphins. Because endorphins play a powerful role in creating that sense of wellbeing and contentedness that is so important in the process of social bonding, it is not hard to see how singing and dancing might have functioned as a device to generate that sense of belonging, or groupishness, that is so fundamental to the coherence of small human communities the world over. There is nothing like a ceilidh (the word means literally 'a visiting' in Gaelic) to bring people together.

This doesn't mean to say that Darwin was wrong, of course. There is every reason why sexual selection should have exploited for its own ulterior purposes the skills and emotions involved in producing music that had evolved for some entirely different purpose. Evolution is very good at doing that, and there are many examples of just this in the animal world. But at root, music's real origins and function probably lie in bonding social groups. And herein probably lay the origins of language itself.

Chapter 7

Why Gossip is Good for You

Why is it that we are so fascinated by what other folks get up to? Why should we find tittle-tattle about the private lives of minor celebrities, royalty, politicians and even each other of such overwhelming interest that it can drive the starving children of Darfur or the war-ravaged cities of Somalia and Iraq off the front pages of even the most sedate of newspapers? The reason is very simple: gossip makes the world go round.

Men talk, women gossip . . .

So how much time did you waste yesterday wittering away nineteen to the dozen? I'll wager it was getting on for a quarter of your entire day. And what came of it all? Probably not a lot, you might say. But it wasn't totally frivolous. It's an odd thing, this language business: we find it intensely embarrassing to remain silent in company. We cast around desperately for something to say, however meaningless. Um . . . do you come here often?

So why do we do it?

One answer is that language is just a form of grooming. For monkeys and apes, grooming is less a matter of hygiene and more an expression of commitment. Its sense is more that of: 'I'd rather be here grooming with you than over there with Jennifer.' We still do a great deal of mutual mauling of this kind, of course. It's an essential feature of all intimate relationships. Parents and offspring, lovers, friends – all are willing to spend hours stroking, touching, leafing through hair. Physical contact, in short, is an essential part of the rhythm of social life.

To this, we humans add language. It's a kind of grooming at a distance and, in many ways, serves much the same kind of purpose. It allows us to make that all-important statement about commitment: 'I find you interesting enough to waste time talking to.' Forget all that highfalutin' nonsense about Shakespeare and Goethe. Real conversations in the everyday world are simply plain honest grooming.

Of course, language allows us to go one step beyond mere signals of commitment. It allows us to exchange information. Monkeys and apes are restricted to direct observation when it comes to learning about who might make a good friend and who is unreliable, or who is going out with whom. But we can learn about these things at second and third hand, and that greatly extends our circle of social knowledge.

Take a listen to the conversation next to you. It will soon become clear that most of our conversations are concerned with social doings. Sometimes our own, sometimes other people's. It's the Harry-met-Sally-met-Susan syndrome.

But nothing comes for free in evolution. Being able to exchange information on who-is-doing-what-with-whom inevitably allows us to use language for more nefarious purposes. In short, advertising should properly be accorded the title of the oldest profession. We are past masters of it. If you don't believe me, listen more closely to that conversation.

There is, however, a curious asymmetry in the conversations of men and women. Harry, it seems, likes to talk about Harry, but Sally talks about Susan. Ah, you say, everyone's stereotypes confirmed. Well, yes and no. There's no smoke without fire, of course. But the really interesting question is why it should be like this.

Men and women's preferred conversation topics are often radically different because they are playing rather different games. Listen carefully to what they actually say, and you soon realise that women's conversations are primarily geared to servicing their social networks, building and maintaining a complex web of relationships in a social world that is forever in flux. Keeping up to date on everyone's doings is as important as the implicit suggestion that you are enough a member of the in-group to be worth talking to. This is not tittle-tattle. It's the very hub of the social merry-go-round, the foundation on which society itself is built.

In contrast, men's conversations seem to be geared as much to advertising as anything else. They talk about themselves or they talk about things they claim to know a lot about. It's a kind of vocal form of the peacock's tail. Male peacocks hang about on their mating territories and display their brilliant tails whenever a female hoves into view. The peahens wander from one male to another, choosing among the males on the basis of their trains.

Humans, it seems, do all this vocally. Like the peacocks that suddenly raise their tails when a peahen is near, men switch into advertising mode when women are present. Have a listen to the same man when he is talking only to other men and compare it with what he talks about when women are present. When there are women present, his conversational style changes dramatically. It becomes more showy, more designed to stimulate laughter as a response. But, in addition, you'll find that technical topics and other forms of 'knowledge' become more intrusive. It's competitive and it's a manifesto. Politics is the name of the game. Language is indeed a many-splendoured thing.

Motherese has so much to answer for

The American anthropologist Dean Falk has suggested that language might have come about through mothers singing to their babies. That peculiar form of speech known as *motherese* which women (in particular) seem to use so naturally when talking to infants has many of the hallmarks of music – a simple rhythmicity, a strikingly exaggerated sing-song intonation that can rise and fall two whole octaves, and a pitch that is significantly higher than normal speech. Next time you overhear a mother talking to her baby, listen closely. You'll be listening to distant echoes of the past. Oh, and don't forget to watch the baby. This unique form of music is very calming for it, and babies seem to find it very attractive and soothing. It stimulates smiling. It's the magic of endorphins again, and their role in bonding.

But motherese has much more important effects than just calming baby. It can dramatically affect the speed with which a baby reaches its developmental milestones. Marilee Monnot, then a postgraduate student in biological anthropology at Cambridge University, observed fifty-two mothers and their newborn babies during the first year of the baby's life. She found that those mothers who used more motherese had babies that grew faster and reached the early developmental milestones (like smiling) more quickly than those who used less. That's quite scary.

Monkey and ape mothers do not croon to their babies. They don't even rock them. It seems to be something that is peculiar to humans. Nonetheless, it's not hard to see how motherese might have got going, though exactly when that might have happened is a tad more difficult to say. If humming soothes baby, and a less fractious baby is more healthy, then there is likely to have been very considerable selection pressure on mothers to do this kind of thing. But why us humans, and not, say, our great ape cousins? The answer surely has something to do with the fact that human babies are born around a whole year premature compared to what we would expect for an ape or monkey of our brain size (I'll have more to say on this later). By comparison, ape babies can pretty much look after themselves. Human babies need an awful lot more attention, and don't really get to the same stage of development as a newborn chimpanzee baby until they reach their first birthday. Since a whole lot more work has to be done by the human baby's long-suffering parent, a mechanism that quietens and soothes a fractious baby must have been all the more necessary in our lineage.

If so, then perhaps this gives us a clue as to when it might have evolved. If it was a response to the radical change in birthing pattern that resulted from the last big upward shift in brain size, then we can perhaps point to the appearance of archaic humans around half a million years ago. This might well have coincided with the origins of music. Motherese might have been the precursor of music, or it might have been the stepping stone between music and language.

Motherese isn't really language. Although it often does consist of words, it doesn't have to. Often, it is just nonsense syllables. It shares much with nursery rhymes – rhythm, rhyme and alliteration. *Hickory, dickory, dock* . . . That in itself suggests that it long predates the evolution of language. It is all so much more like wordless singing, or humming – pure music. In this, it shares a great deal with sea shanties. And it also shares a great deal in common with that most extraordinary and unique form of vocal music, the waulking songs (*òrain luaidh* in Gaelic) of the women of the Outer Hebrides. Part just nonsense syllables, part witty – often raunchy – reflections on lives coloured by poverty and hard work, and not infrequently by tragedy, these extraordinary songs have been sung for centuries by the women as they stretch and soften the newly woven tweed round a kitchen table. Passed down by word of mouth from one generation to the next, they are a remarkable and unique tradition. I wonder if they don't represent the very first kinds of situations when language was used – by women around the campfire, or out foraging for fruits and tubers. There is something about synchronised singing that seems especially good at triggering the release of endorphins: many voices make light work.

The importance of a good gossip

In the end, of course, language evolved to allow us to integrate a large number of social relationships. And the way it does this is by allowing us to exchange information about other individuals who are not present. In other words, by talking to one person, we can find out a great deal about how other individuals are likely to behave, how we should react to them when we actually meet them and what kinds of relationships they have with third parties. All these things allow us to co-ordinate our social relationships within a group more effectively. And this is likely to be especially important in the large, dispersed groups that are characteristic of modern humans.

This would explain our fascination for social gossip in the newspapers, and why gossip about relationships accounts for an overwhelming proportion of human conversations. Even conversations in such august places as university coffee rooms tend to swing back and forth between academic issues and gossip about individuals. To get some idea of how important gossip is, we monitored conversations in a university refectory, scoring the topic at thirty-second intervals. Social relationships and personal experiences accounted for about seventy per cent of conversation time. About half of this was devoted to the relationships or experiences of third parties (people not present).

But since males tend to talk more about their own relationships and experiences, whereas females tend to talk most about other people's, this might suggest that language evolved in the context of social bonding between females. Most anthropologists have assumed that it evolved in the context of male–male relationships, during hunting for example. The suggestion that female–female bonding, based on knowledge of the relationships of other individuals, was more important fits much better with views about the structure of nonhuman primate societies where relationships between females are all-important.

That conversations allow us to exchange information about people who are not present is vitally important. It allows us to teach others how to relate to individuals they have never seen before, or to handle difficult situations before they have to face them. Combined with the fact that language also makes it easy to categorise people into types, we can learn how to relate to classes of individuals rather than being restricted to single individuals as primates are when grooming. We can agree to give types of individuals special markers, such as dog collars, white lab coats or large blue helmets, which allow us to behave appropriately towards them even though we have never met before. Without that knowledge, it would take us days to work out the basis of a relationship.

Classifications and social conventions allow us to broaden the network of social relationships by making networks of networks, and this in turn allows us to create very large groups indeed. Of course, the level of the relationship is necessarily rather crude but at least it allows us to avoid major social faux pas at the more superficial levels of interaction when we first meet someone we don't know personally. Significantly, when it comes to really intense relationships that are especially important to us, we invariably abandon language and revert to that old-fashioned primate form of direct interaction – mutual mauling.

What we seem to have here, then, is a theory for the evolution of language that also seems to account for a number of other facets of human behaviour. It explains why gossip about other people is so fascinating; it explains why human societies are so often hierarchical; it predicts the small size of conversation groups; it meshes well with our general understanding of why primates have larger brains than other mammals; and it agrees with the general view that language only evolved with the appearance of modern humans, *Homo sapiens*.

What it does not explain, of course, is why our ancestors should have needed to live in groups of about 150. It is unlikely that this has anything to do with defence against predators (the main reason why most nonhuman primates live in groups) because human groups far exceed the sizes of all other primate groups. But it might have something to do with the management or defence of resources, particularly dispersed resources such as water holes that nomadic hunter-gatherers might have had to depend on at certain times of the year.

Now tell me another story

Language is also crucial for one of our most peculiar activities – story-telling. It is something that humans all around the world do and love, and surely have done ever since time immemorial. It is not just a bit of old gossip, for stories told around the campfire are imbued with ritual and often have a very formal structure. Many are incredibly old, such as the great Hindu epic the Mahabharata, written around two thousand years ago, or the stories contained in the books of the Old Testament or the

Bhagavadgita that were composed some five hundred years earlier, just a few centuries after Homer's great epic poems the *Iliad* and the *Odyssey*. Some of the stories told by Australian Aboriginals living along the south coast of the continent appear to be even more ancient: they are said to contain surprisingly accurate descriptions of the landscape on the sea floor of the Bass Strait that separates Tasmania from the Australian mainland – a land surface that was last exposed as dry land during the Ice Age that ended twelve thousand years ago.

So why should we be so fond of stories?

Well, for one thing, many such stories are origins stories – they tell us where we came from, and how we came to be the way we are. They tell us about community, they create a sense of belonging for us.

Shared knowledge itself is a good marker of community membership. That you know immediately what I mean when I observe that silly mid-on dropped the catch inexorably marks us out as belonging to the same community, the community of those who play or follow cricket. By virtue of that simple fact, we can be sure that we share enough in common to be willing to exchange favours should that ever be necessary. We have a common world view, and by implication subscribe to a common set of rules about how one should behave. It probably reflects the fact that, in our deep past, people who shared such knowledge lived together, and were almost certainly related to each other. So, discovering that we share esoteric knowledge still seems to create an instant bond between us, sets us apart from the rest of the common herd. That may be one reason why we are so fond of creating technical jargon – it sets us apart as special, a shadowy secret cabal that knows the innermost secrets of the universe. There's nothing like a good secret.

But while there is something deeply engrossing about a good story well told, there is perhaps nothing quite so captivating as a story told around the campfire at night. We seem to be especially fond of story-telling at night, and there can hardly be a culture around the world in which this is not true. But why should darkness make stories seem so much more vivid?

It's not enough to say that the evening round the campfire is the only time you have for relaxation – the day's work is done, there is nothing more to do, so idle chatter can fill the time before bed. That's not really a convincing explanation because, if there really was nothing useful to do, we could just as easily go to sleep as soon as it gets dark just like all the other sensible monkeys and apes. But we don't: we stay up and chatter. What's more, it's a peculiarly social time, the time when we prefer to invite guests for dinner – even at the weekend when the days are presumably uncluttered by work and we could easily have invited them for breakfast, lunch or tea, it's still dinner that we prefer. Of course, we can – and sometimes do – sit around the campfire of an evening doing useful chores like making or repairing clothing or hunting equipment. Yet we still tell stories while we do these things.

Perhaps it has more to do with psychology and the ambience. Perhaps story-tellers find it easier to play with our emotions in the dark, and we rise to that precisely because we get more of a kick out of it. Perhaps it is because many such stories are about mythical creatures, and daylight casts too much of a cold dose of reality on them to make them believable. Such stories may need the uncertainty of the dark, when we feel vulnerable because danger – whether natural predators or human muggers – can too easily get within our 'escape distance' (the distance at which we can still evade a predator once we have detected it). Perhaps it is just easier for a skilled storyteller to work on the audience's emotions at night.

Chapter 8

Scars of Evolution

We are, as Darwin reminded us in the *Descent of Man*, the product of a long evolutionary history. We still bear the scars of that evolutionary history today. Some of those scars – like our particular skin colour – are relicts of a surprisingly recent evolutionary history and date only to within the last few tens of thousands of years. Most of these are recent genetic mutations triggered by the great migrations out of Africa that resulted in modern humans peopling the whole planet. Others, however, are older, dating back to earlier species in the lineage that led to modern humans. One of these is the fact that, unlike all other primates, we give birth to very premature babies – one of the consequences of which seems to have been the need to persuade males to get involved in the business of child-rearing in a way that is very rare among mammals. And speaking of babies reminds me of milk – that special stuff invented by mammals on which to nourish their babies.

Our love/hate relationship with milk

If, like me, you are of a certain age, you will remember that morning ritual at school when your exodus into the playground was delayed for precious minutes by the arrival of a little bottle . . . Love it or hate it, it was the moment when the milk turned up on your desk – just a shade short of ice pops in winter, curdled into something close to cheese in summer. Be grateful for a moment: most of us downed the stuff and occasionally even enjoyed it. But are you aware that, if you did down it with even a modest amount of pleasure, you are actually among the privileged few? Did you know that most of the people in the world cannot drink milk without becoming ill?

That's not because they have some serious medical condition. It's because we milk-drinkers are the aberrant ones. We all have a unique mutation that is only found in a small minority of modern humans – a mutation for the enzyme lactase that allows us to digest lactose, one of the main sugars in milk. Of course, all humans can digest milk as babies. But for most of the world's population, the lactase gene that allows us all to do this is switched off at weaning; after that, milk, and milk products, become indigestible and consuming them can have the most unpleasant, even fatal, consequences.

It was only during the Second World War that we became aware of this. Milk had been such a central part of European culture that no one gave it a second thought. It was, after all, very good for you – rich in proteins and energy, and loads of calcium for growing bones. So when the US government wanted to build up the health of their more deprived populations, someone had the obvious answer – milk, and lots of it. To everyone's consternation, it had just the opposite effect in the black communities. Children started going down with diarrhoea and losing weight. More by luck than judgement, not many of them actually died – which, had this well-intentioned intervention gone on much longer, might easily have been the inevitable outcome.

Puzzled by this, the scientists set to work to figure out what had gone wrong. Eventually, it transpired that the ability to digest fresh milk post-weaning is a peculiarity of Caucasian peoples (and particularly people of northern European extraction at that), plus a few cattle-keeping people along the southern fringes of the Sahara. Almost everyone else in the world avoids milk like the plague – or at best consumes milk only in highly processed forms like yoghurts or cheese, or better still by cooking it to death first.

Which is also why sending powdered milk to Africa during famines is probably not the smartest thing to do. Doling out large quantities in such situations is the best way of making a bad situation worse. It can place the lives of babies, already weakened by famine, in even greater peril.

So how did this odd state of affairs come to be?

The answer, it turns out, has to do with the fact that – as northerners know only too well – the sun gets steadily weaker as you head into higher latitudes. The problem is that human skin synthesises vitamin D as a part of a reaction to ultraviolet (UV) light, and this is the only way we can acquire this essential vitamin naturally. Calcium is involved in this process, so being able to consume lots of extra calcium helps the body to synthesise vitamin D more effectively in the watery sunlight of the north. Having light-coloured skin helps enormously because it allows more UV light to penetrate the surface. The dark skins of the tropics – caused by a layer of dense melanin cells under the skin surface – are designed explicitly to reduce the amount of harmful UV light that would otherwise penetrate. More of this in a moment.

Tolerance of lactose involves just a single gene mutation – not in the form of a novel gene as such, but rather as a fault in the mechanism that would normally switch off the gene that codes for lactase, which would normally happen at weaning. So the genetic change required was very modest. But the genetic change on its own was not enough: it also required a cultural change to encourage the keeping of dairy animals and an enthusiasm for drinking calcium-rich milk.

Northern latitudes also have another problem that is not encountered in the tropics – they are much more seasonal. In the tropics, the growing season often extends virtually throughout the year, and several successive crops can often be sown and reaped each year. As you go further north and the climate gets more seasonal, the growing season becomes so short that it imposes long stretches of the year when things can be pretty tough. Having milk to fall back on means that you don't have to slaughter your entire herd to survive the winter. Domestic animals become a walking larder.

And this might explain why lactose tolerance also occurs among the milk-drinking, cattle-keeping people like the Fulani who live in the Sahel, the arid zone along the southern border of the Sahara. This is an area that has always been, and very much still is, prone to famine. How useful to be able to resort to a renewable form of food-on-the-hoof when times get bad!

Skin deep

Mention of skin colour raises that hoary old question of why people who live in the tropics tend to be darker-skinned than those who live at higher latitudes. I mentioned that this has something to do with keeping harmful sunrays out of the skin. A recent study by Nina Jablonski and George Chaplin of the California Academy of Sciences has done much to clarify this.

They were able to show that seventy-seven per cent of the variation in skin colour in northern-hemisphere peoples, and seventy per cent of that in southern-hemisphere peoples, correlates with the level of ultraviolet radiation (UVR) – the component of sunlight that is so damaging to skin cells and, as light-skinned folk are now so often warned when heading for the beach, a major cause of skin cancer. UVR levels decline as you go progressively further north or south of the equator because the curvature of the earth means that, with the sun positioned more or less above the equator, there is more air mass for the sun's rays to pass through. Since sunlight is absorbed by the air, less UV radiation reaches the earth's surface as you get nearer to the poles.

However, UVR levels do not correlate perfectly with latitude. High-altitude areas that lie at mid-latitudes, like Tibet and the Andes plateau of South America, have high UV levels because there is less air mass above them to absorb the harmful rays. Similarly, local cloud cover has an effect because water vapour in the atmosphere helps to filter out UVR. The Atacama desert in Chile and the deserts of the southwestern USA and the Horn of Africa have unexpectedly high UVR levels for their latitude because they are so dry and, unlike the lucky folk in the British Isles, lack a nice layer of clouds above them.

Jablonski and Chaplin argue that the evolutionary origins of this variation in skin colour actually have less to do with skin cancer than with a trade-off between the competing benefits associated with two different vitamins. One is the extent to which sunlight breaks down vitamin B (folic acid). The melanistic cells that produce dark skin tones (and suntans in pale-skinned Europeans) protect vitamin B in the skin from sunlight. Like all primates, we don't synthesise vitamin B, but instead have to acquire it by eating the meat of animals that do. Protection against excessive sunlight thus helps to reduce the need to worry about the lack of vitamin B in our diet.

The converse, however, holds with vitamin D, a vitamin that is important in calcium absorption (and hence, strong bones). We can synthesise vitamin D for ourselves, thanks to a reaction between sunlight and skin cells. However, when light levels are low, as they are at high latitudes, people with dark skins cannot synthesise enough vitamin D. Albino African children in South Africa, for example, require less dietary vitamin D supplement than children with normal dark African skin colour. Hence, lighter skin colours have evolved in more northerly populations. (Since there isn't much land surface outside the tropics in the southern hemisphere, there isn't a native white southern race like there is in the north. However, it is far enough south that the ancestral inhabitants of southern Africa – the San Bushmen – have a lighter-coloured, more coppery skin than the dark-skinned Zulus who arrived in southern Africa only a few hundred years ago.)

One surprising observation that supports this explanation is the fact that women and babies typically have lighter skins than adult men in all human races, including among Africans. Women have a particular need for calcium and vitamin D during pregnancy and lactation – in traditional societies, after all, women spend much of their adult lives in one or the other of these two states. Having a higher capacity to synthesise vitamin D is thus beneficial for the women.

Despite the neatness of this explanation, we are left with several puzzles – why is the relationship between skin colour and latitude/UVR quite a bit stronger in northern-hemisphere peoples than it is in southern-hemisphere peoples? And why, given the importance of vitamins, is the relationship not perfect?

As it turns out, the answer to both questions has to do with a combination of history and culture. The biologist and polymath Jared Diamond has pointed out that many populations whose skin colour is 'out of place' are peoples whose ancestors undertook lengthy migrations within recent historical times. Thus, the dark skins of the Bantu peoples of southern Africa reflect the fact that their ancestors arrived in southern Africa from a west African homeland near the equator only within the last few hundred years. Similarly, the rather light skins of many southeast Asians (Filipinos, Cambodians, Vietnamese) reflect the fact that their ancestors migrated from a homeland in southern China only about two thousand years ago. All the descendants of the original inhabitants of these countries (often collectively known as 'hill tribes' and 'negritos') have much darker skin colours.

One illuminating exception is provided by the Eskimo,

who have somewhat darker skins than we would expect for a people who live so far to the north. The explanation is that they rely heavily on marine mammals like seals and polar bears as a source of food. These species have livers that are especially rich in vitamin D, and liver is much favoured as a food by Eskimo peoples. Since this helps take care of their vitamin D problem, it allows the interests of vitamin B to take precedence and select for darker-coloured skin – which is why the Eskimo have their classic coppery-coloured skin.

For most of us, skin colour is a function of where our recent ancestors have lived. Even so, the pace of change can be astonishingly fast in evolutionary terms. The ancestors of modern Europeans have occupied the more northerly parts of Europe only since the end of the last Ice Age, a mere ten thousand years ago. The blondness of Scandinavians probably has a very short history.

Why giving birth is such a pain

Babies have their own appeal, and never more so than to their doting parents and grandparents. It's probably just as well, since human babies are born wildly premature. In mammals as a whole, the length of gestation is dictated by the size of the brain. It seems that brain tissue can only be laid down at a set rate, so if you want to grow a big brain, you can only do it by growing your brain for longer. Species with large brains thus typically have long gestation periods. In effect, it is the babies who decide when they are ready to be born – a theory in biology that has come to be known as 'the baby in the driving seat'.

The problem for us humans is the sheer size of our

brains. On the basis of the pattern we find in the rest of the mammals, we humans ought to have a gestation of twenty-one months. Yet as we all know, it is actually just nine months. The reason is very simple. Several million years before our ancestors decided it might be a good idea to have such big brains, they thought it was an even better idea to walk upright. This led to the evolution of our very distinctive bowl-shaped pelvis, quite different from the rather elongated pelvis of all the other monkeys and apes. This bowl-shaped pelvis provided a much better base on which to balance the trunk and head, especially once that big bulging brain came along. The modern human pelvis has been with us for the better part of the last two million years, ever since the first members of our genus, *Homo erectus*, developed their striding walk and the capacity to migrate over long distances.

The problem is that, as invariably happens in evolution, it is impossible to get a perfect engineering design. One of the sacrifices we have had to put up with to get the benefits of long-distance striding has been a weak lower back. Evolutionary processes could, of course, have solved the problem by making the lower backbones out of cast iron, or perhaps bone ones of massive proportions, but that would have added measurably to the weight we have to carry around, and would have made our lower back much less flexible. That flexible spine is a trait of enormous value to our walking pattern – and of major significance to cricketers who fancy themselves as fast bowlers – and so, by definition, to our many ancestors who made their way in the world by spearing wild animals for meat. What we have is a classic Heath Robinson evolutionary compromise – a consequence of trying to get the best of two worlds that have conflicting interests. The painful result is a lower back that is still prone to 'go'.

Then, when their descendants decided to increase the size of their brains dramatically several million years later, they hit a bit of a problem: the bowl-shaped pelvis had dramatically narrowed the birth canal. Since it is the size of baby's brain that is the limiting factor, the result was . . . well, an eye-watering problem.

At this point, the options were rather limited. Of course, we could have backtracked rapidly and given up this stupid idea of having bigger brains – who needs brains anyway, for heaven's sake? But that would have meant staying put in our evolutionary niche. Since, thanks to climate change, the world was altering dramatically around this time, staying put would have meant becoming ecologically more embattled, just like the other great apes whose terminal decline towards extinction was already well in motion by then. To survive, we had to change and adapt to new ecological niches. Big brains were the key to that, and without them those kinds of changes wouldn't have been possible. So something dramatic was called for.

The inspired solution our ancestors eventually came up with was to reduce dramatically the length of time for which the mother carries the baby before birth . . . from the twenty-one months we ought to have, to the nine we finally settled for. But this came at a cost: giving birth to a baby whose brain is only half-developed means a very vulnerable baby. Whereas monkey and ape babies are active and busy within a few hours or at most days of birth, human babies take a full year – the missing twelve months – to reach that stage.

Compared to monkey and ape babies, human babies are on the very edge of survival even when they are full-term. This is why they really do struggle when they are born prematurely. Research in the last decade or so has found that premature babies suffer disproportionately high frequencies of developmental difficulties, including poorer academic performance and more physical problems later in life. This is not, of course, to say that every single one of them does, but rather that the risks are just much greater.

So it is that for the first year of life, normal human babies are basically just lumps of flesh and bone that need an awful lot of TLC. And since TLC is hard work for parents, babies had better have winning ways and lots of baby-appeal. And that raises a whole new host of problems. One of these is the fact that, from the mother's point of view, it pays to have your man on side. But if the baby is not his, that can create – shall we say – difficulties. At this point, you have two choices: you can make sure the baby really looks just like dad, warts and all, or you can make it look like no dads. The first is fine so long as dad always really is dad. But if dad – shall we say with even greater delicacy – isn't always dad, maybe the second is the better option. And that, it seems, is what humans have done. Human babies by and large look much more like each other than adults do. So much so, in fact, that all babies have blue eyes to begin with, and only change into brown or green later. It helps keep dad guessing.

But anxious not to leave anything to chance, we back this up with a bit of psychology. Next time you are around a newborn baby – probably best not yours – listen to what people say about it. A study by Martin Daly and Sandra

Wilson of McMaster University in Canada found that both the mother and her parents make great efforts to emphasise how much the baby looks like dad as soon as he comes into the room. 'Hasn't he got your eyes/nose/forehead/chin . . .' And this wasn't just a Canadian or European thing: similar results were reported from another study in Mexico. Now, excuse me . . . but nothing on baby's face looks anything like any of its progenitors' equivalent bits. It's not meant to. Still, it does provide a lot of scope for leverage to persuade dad that he had better get his sleeves rolled up. Just as well, probably.

Just how complicated can sex get?

I'll confess straight away that I am fascinated with sex. Never in the course of biological evolution has anything more complicated ever evolved. And I don't just mean the complications of the relationships that emerge from it. I mean biologically. I'll bet you think that sex is just about X and Y chromosomes. At least, that's what you were probably taught in school biology lessons. And, up to a point, it's true: we are bog-standard mammals, and our sex is determined by the chance event of whether we inherited an X or a Y chromosome from our father to pair up with the X provided by mum. XX gives a girl, XY a boy. Simple, isn't it? Well, yes, up to a point. But in fact, it's a bit more complicated, even in humans. Your sex chromosomes are only part of the story. You may have an XY pair, but you need not have turned out to be a boy.

In fact, you only get to be male if a whole series of things fall into place at the right time – otherwise you will be female, whatever your sex chromosomes. One of these key events is known as the 'race to be male'. The foetus lays down a particular type of fat cell early on, and it requires a specific density of these to switch an XY-chromosome foetus over from its default female body form to a male one. The right density of fat cells triggers the release of testosterone that switches the foetus's brain over into a male brain, and this sets in motion the conversion of all the other bits that matter.

In fact, even chromosomal sex can get pretty confusing. Accidents of genetics can result in any number of possible combinations – Xo (one X chromosome and nothing else), XXY, XXYY, XXXYY, XYY (the so-called 'super male'). The only one you can't have is Yo (no X chromosome): the Y chromosome is tiny and only a very small segment of its DNA has any function, and that is associated with the business of changing the default female form into the male one. But if you don't have the female bit to start with . . . curtains, I'm afraid. However, that said, most of this bewildering array of chromosome types are associated with fairly serious disabilities and abnormalities, so their product is often distressing. Fortunately, most of them are rare.

But things begin to look even odder when you look beyond us mammals. Birds, butterflies and amphibians do it the other way around. In birds, it is the XY sex that lays eggs and the XX sex that has gaudy plumage, sings songs and rushes around defending its territory. To avoid confusion, the bird people usually refer to these as W and Z chromosomes, rather than X and Y, but that doesn't hide the fact that they are the mirror image of us mammals. What this tells us is that it's an accident of history which way things turned out: there is no 'natural' way of doing sex.

And it gets worse. In turtles and crocodiles, your sex depends on the temperature of the nest in which you were incubated as an egg. In crocodiles, warm temperatures produce males, cooler ones females, but in turtles, it's the reverse. Famously in bees, females have two sets of chromosomes, but males have only one (because they arise from unfertilised eggs). In many of the small coral reef fish like wrasses, it depends on social circumstance. Everyone begins life as a female, but if there is no male, the dominant female in any community undergoes a rapid metamorphosis and miraculously turns into a male before your very eyes. When she – or should it be he? – dies, the cycle starts again and the current dominant female changes sex and becomes the breeding male. I guess that gives a new meaning to the phrase 'change of life'.

But of all the bizarre and weird ways in which species produce two sexes, perhaps the first prize goes to the humble bonellia worm, a ten-centimetre-long member of an obscure worm family found in the Mediterranean sea. All bonellia begin life as tiny flake-like larvae, floating free. Those that happen to get attached to rocks or other substrates turn into females; those that get eaten by a female before finding somewhere to attach migrate down into the female's uterus and turn into males. They then spend the rest of their lives in the safe confines of the female's interior – which they may share with up to twenty other males.

Sex is fascinating – I rest my case.

Chapter 9

Who'd Mess with Evolution?

The medical profession has a great deal to answer for. For millennia, it has held us in the palm of its hand because of our desperation to avoid the inevitable consequences of our biology – disease, disability and death. As medical science has become more sophisticated, it has seemed able to perform miracles. But one of the problems is that, more often than not, those miracles pander to our short-term desires rather than to what is best for us in the long run. We are desperate for short-term cures that solve a problem now, but we ignore the fact that doing so may create bigger problems for us in the future.

It seems that we never learn. During the 1950s, DDT and penicillin seemed to be the wonder drugs of the century: we could cure anything from malaria to infections that had previously killed hundreds of thousands of children and adults every year. So we liberally sprinkled DDT over tropical habitats, and dosed ourselves and our animals on penicillin. But natural selection, the engine of evolution, soon undermined all this good work. Within just a few decades, we had successfully, if unintentionally, bred DDT-resistant mosquitoes, penicillin-resistant bacteria, MRSA and a string of other horrors that have made our original problems seem like kids' play. The moral is that it isn't always sensible to try to interfere with evolution – especially when, like most members of the medical and pharmaceutical professions, you don't really understand the principles of Darwin's theory of evolution by natural selection.

Medicine isn't always good for you

If you have the impression that we are being progressively swamped by new and more troublesome diseases, it's now official: we are. An analysis of 335 major new disease outbreaks that have occurred since 1940 has shown that the frequency of new diseases has increased steadily with time. The number of new diseases that have hit us each decade has increased three- to four-fold in the last half-century alone. Among the more familiar ones are the likes of MRSA and the various new strains of 'superbug' that are resistant to antibiotics, SARS, HIV, and drug-resistant strains of malaria. Given that malaria was already one of the great killers – it afflicts 515 million new people each year, and kills between one and two million of them, mostly children – the prospect of worse to come is not appealing.

Around fifty-five per cent of these new diseases are bacterial in origin, with many fewer than had previously been expected being due to viruses or prions (the most familiar of which is 'mad cow' disease). Many are associated with the appearance of drug-resistant forms of old diseases rather than something entirely new . . . a stark and terrifying reminder of the speed with which micro-organisms can evolve when challenged – and of the fact that we have been hoist upon our own petard through the over-liberal, and invariably careless, use of antibiotics and other drugs.

It seems that sixty per cent of these new disease out-breaks are caused by zoonotic pathogens – pathogens that we have caught from animals – and seventy per cent of those have come from wildlife. The notorious Ebola, HIV, SARS and the Nipah virus (a pathogen from fruitbats which appeared in Malaysian pig farms in 1999 and resulted in 105 human deaths) are all cases of pathogens that jumped the species barrier from their natural animal hosts to humans.

This is not entirely new, of course. Many of our more familiar diseases – often ones that, in the past, have caused very high mortality rates – had their origins thousands of years ago in domestic animals that our ancestors decided to have living with them or were brought into our houses by rodents of one kind or another. Chickenpox, cowpox (and its close relative smallpox), measles, rabies, Lassa fever and haemorrhagic fever all have their origins in pre-history, thanks to our having had too close an association with their respective animal hosts.

The tropics have been notorious breeding grounds for most of these historic diseases, and it has long been recog-nised that the tropics are among the least healthy places to live – unless one happens to descend from a racial group that has evolved some kind of immunity over time. Examples of the latter include the well-known case of sickle cell anaemia among west African Bantu peoples. Sickle cell is a recessive allele that confers significant resist-ance to the malaria parasite, but when a recessive allele is inherited from both parents, the result is an excruciatingly painful condition whose sufferers rarely make it beyond their teens.

The fact that dread diseases are more common in the tropics than at higher latitudes may in part explain a curious feature of how languages are distributed: near the equator, language densities are much higher, and language communities (the number of people speaking a given language) very much smaller, than they are at higher latitudes. One explanation for this might be that it is a culturally evolved strategy to reduce the risk of cross-infection in areas where pathogens are more densely concentrated. Language barriers significantly reduce the opportunities for contact between different populations, thus minimising the risk of contamination. Creating smaller, more inward-looking, xenophobic societies may thus help to reduce exposure to diseases to which one has no natural immunity. It turns out that religion has a similar distribution: Randy Thornhill and his colleagues at the University of New Mexico found that people living in areas with high parasite loads (mostly those in the tropics) are much more religious than those living in areas with low parasite loads (mainly those at high latitudes).

Nonetheless, despite the fact that many new diseases seem to have their origin in the tropics, it is often in the subtropics that the major outbreaks occur. This seems to be explained by the fact that human population density is the single most important factor relating to disease outbreaks. In part, that reflects the historical development of the more successful economies of Eurasia and North America, creating denser populations of susceptible individuals. In addition, of course, language communities are significantly larger outside the tropics, thus facilitating mutual intercourse (in all senses of the word) between larger numbers of individuals.

In the end, however, the high proportion of these new diseases that have their origins in wildlife (so-called zoonoses) means that the single best predictor of where these diseases originate is local wildlife biodiversity. And that is a tropical issue. What should concern us is the fact that most of these biodiversity hotspots are in developing countries in Africa, Asia and Central America – the ones where the investment in disease monitoring and control is least well developed. It raises the question as to whether we in the developed world are investing our resources as wisely as we should because, once such diseases have moved from the developing to the developed world, they are invariably much more difficult to deal with. It's a very good reason for putting more money into the developing nations.

Curse morning sickness

If you suffered with morning sickness in early pregnancy, it may be little consolation to know that you are not alone: four out of five mothers-to-be experience vomiting or food aversions in the first three months. The medics, as usual, have tended to see only the symptoms and settle for offering palliatives of mostly questionable value – thalidomide, which blighted so many lives in the 1960s, was just one of the least sensible: it stopped the symptoms of morning sickness, but no one really took the trouble to look beyond that. In the medics' considered view, morning sickness is just an unfortunate side effect of the hormonal changes that happen during pregnancy, so there is every good reason to get rid of it. But evolution doesn't often produce things that are mere side effects. So why on earth should we experience such awful side effects from what is, after all, a perfectly natural process of everyday life?

In fact, it seems that morning sickness might actually be good for you – or at least, for your baby. Women who experience nausea in the first trimester of pregnancy have greatly reduced chances of losing the baby by spontaneous abortion, and are likely to give birth to bigger, bonnier bairns. This has prompted evolutionary biologists to ask why this should be. One suggestion is that it is the outcome of a tussle between baby and mother over what the mother should eat. The argument is very simple. We eat lots of things that are mildly toxic, sometimes even downright poisonous, often because they taste good or give us a kick of one kind or another. These include things like alcohol, coffee, chilli, pepper and even broccoli. Many of these are carcinogens (cause cancers) if taken in large enough doses, and not a few are teratogens – substances that cause abnormalities in developing babies if ingested too often during pregnancy.

Adults can tolerate these poisons because the small doses we eat are diluted when dispersed around our relatively large bodies. But foetuses are tiny, and to receive even a small dose of one of these via the mother can have very adverse effects. In effect, morning sickness is the baby's way of trying to prevent the mother from eating too much of what is not especially good for baby.

An alternative suggestion has been that the vomiting associated with morning sickness gets rid of harmful bacteria ingested with foods that are prone to going off. Adults can usually cope with a little rotting meat in small doses

– it may cause an upset stomach, or at worse a spot of diarrhoea, but that passes pretty quickly. But, once again, what may be a tolerable dose for mum may be just too much for baby. The obvious candidates are meats and dairy products.

In a recently published study, Craig Roberts and Gillian Pepper of the University of Liverpool looked at the frequency of morning sickness across the world in relation to the kinds of diets typically eaten. They found that morning sickness frequencies were indeed correlated with the frequency with which stimulants (such as coffee) and alcohol were consumed. However, the frequency of morning sickness was most strongly associated with the amount of meat, animal fats, milk, eggs and seafood eaten, and least with the importance of cereals and pulses in the diet.

This suggests that it might well have been the risk of damaging infections that has played the major role in the evolution of morning sickness. The association between morning sickness and the amount of meat and dairy produce in your diet makes sense if the real problem is to avoid poisons. Meat and dairy produce are, after all, among the most nutritious foods available: they are rich in easily digestible nutrients. Why should one avoid them? The answer can only be the fact that they are prone to being infected with bacteria, and the load that these place on mother and baby may be enough to trigger a spontaneous abortion. Most cereals don't have this problem, so the more cereals you have in your diet, the less trouble you get.

One curious bit of evidence against the poisons hypothesis is the fact that the frequency with which spices are used in food also correlates negatively with morning sickness rates – which is surprising since many spices are well-known carcinogens. However, as every traveller to the Far East knows, a good hot curry kills off everything including all the bacteria inadvertently ingested with your food. Spices, it seems, are good for you. They also happen to be quite good at triggering the release of endorphins – the brain's own painkillers – and these in turn seem to 'tune up' the immune system, thereby making you better able to cope with illness.

So, if you are thinking of getting pregnant, it seems that avoiding meat and dairy produce may be the best way of reducing the risk of morning sickness. That age-old Scots remedy for everything, porridge, is suddenly especially attractive. But maybe you should consider a drop of chilli to spice it up for good measure?

A medical bridge too far?

Pregnancy reminds me that if there is anything that we worry more about than death, it must be not being able to make babies. More anguish, time and money is spent on fertility treatments than anything else except making ourselves smell nicer. And so it was that, in the summer of 2006, thanks to the wonders of science, Patti Farrant became the proud mother of a bouncing baby boy at the age of sixty-two, acquiring at the same time the privilege of being the oldest mother in Britain. But as the latest in a trickle of post-menopausal IVF pregnancies, she raises a more fundamental issue than the mere question of whether grandmothers make good mothers.

We are the product of evolution, and the processes of evolution inevitably instil in us a complex set of motivations and emotions that are designed to serve the principal function of doing evolution's business – ensuring that, as best we can, we each make our contribution to the next generation's gene pool. Because the evolutionary processes are blind to long-term consequences, they operate through emotions that have been tuned over evolutionary timescales to achieve the ends that best serve evolution's interests.

For this reason, we are bedevilled by emotional short-sightedness. It often requires enormous self-control to resist satisfying our cravings in a world where technology can transform craving into actuality. Obvious everyday examples include our tendency to eat too much – and especially too many sugary and fatty foods – to enjoy the momentary pleasures of substances that inevitably harm us in the long run (I need not list these by name . . .), to take risks (of both a sexual and a physical kind) for the thrill of the moment, to over-fish the seas or cut down the forests despite the fact that we all agree that such behaviour is inevitably self-destructive in the long run.

The hardest of these cravings to resist are, surely, those that have to do with our children. Parents are deeply ingrained by evolution to care desperately about their babies. We have to be, otherwise human babies would never survive, given that they are born so prematurely by monkey and ape standards. The problem is – as every parent knows only too well – it doesn't stop with weaning. The need for parents to invest in their children goes on and on and on . . . seemingly for ever.

We tend to forget that successful child-rearing is not just a matter of making sure the little darlings survive childhood. We are an intensely social species, and, from an evolutionary point of view, placing our children advantageously in the adult social world is more important than their mere survival. That involves a great deal of social training in the teenage years, not to mention looking after their economic interests as young adults and providing them with the right kinds of social opportunities, marriage partners or even business breaks. It may begin with finding them godparents; it runs on into finding them jobs with friends or relations, and ends (or so one always fondly hopes) with lavish weddings. And then the grandchildren arrive, and the cycle starts again. Not to put too fine a point on it, it's the first forty years of childcare that are the worst.

The problem is that the wonders of medical science have meant that babies who once would never have survived can now do so. The emotional investment of both parents and medics converge, and a 'can do, should do' culture prevails for what are surely the most understandable of reasons. But is it always in everyone's best interests? In the heat of the moment, parents cannot see beyond the immediacy of their emotions, doctors cannot see beyond the exhilaration of achieving a result against the odds. The pressure is to push the boundaries further and further back, but the consequences down the line get overlooked. This is most serious where babies have more problems than just prematurity: the pressures of coping with the severely disabled all too easily result, during the decades that follow, in intolerable family burdens even for saints. Divorce rates are higher than average, and disabled children are at much greater risk of physical and mental abuse, and even death, when the patience of saints finally cracks under the strain.

Life, and especially growing up, is a risk at the best of times. So is it morally right for those who dabble in this area to take the view that just because it is possible, it should be done? Is it really in our best interests for medical science to be driven by the desperation of our cravings? The lesson of evolution is that, more often than not, the answer is a resounding 'no'.

Boys can be too much of a good thing

Not all evolutionary slaps in the face come from the medical profession's activities, of course. Just as many come from politicians and the social policies they try to impose on us, even if often for the best of political motives. But the consequences of trying to interfere politically with the biological world can be just as problematic. Two decades ago, for example, China worried so much about the population explosion looming over its head that it instituted its now infamous one-child policy: couples were allowed to have only a single child, and any extra conceptions that followed had to be aborted. Draconian as this may sound, it pretty much saved China from demographic disaster. It cut the birth rate overnight, and all but reversed the population growth rate.

However, they had reckoned without the effects that evolution has had on human nature. Lurking unseen in the wings was a completely different demographic disaster. What the government demographers had not anticipated – and demographers in general have never been noted for their understanding of, or even interest in, evolution – was the average couple's overriding preference for boys, especially in rural populations where boys are essential as labour on the farm. The availability of cheap means for sexing babies *in utero* allowed parents to selectively abort girl foetuses.

Now, less than two decades down the line, the hidden timebomb created by an imbalanced sex ratio is beginning to reveal itself. The hundred largest cities in China have a sex ratio of around 125 boys to every hundred girls – against a normal sex ratio at birth of about 108 boys for every hundred girls. Current estimates suggest that there are around eighteen million more men than women of marriageable age in China, and the forecast is that this will rise to thirty-seven million by 2020. This is just a tad ominous, because boys without girls are seriously bad news.

One recent study demonstrated a strong correlation across mainland US states between the divorce rate and the frequency of rape: the significance of this is that many more divorced men than women remarry, and high divorce rates thus result in large numbers of other males being left without partners – and, hence, a large number of excessively frustrated men. And in case you need any more persuasion about the civilising influence of a 'good woman', consider the fact that one of the strongest predictors of recidivism in young male criminals in the UK is whether or not they settle down with a long-term partner after release from prison. Boys without girls are, to be blunt, a menace.

This is not just a recent phenomenon that we can blame on the temptations of the modern world. The Portuguese nobility faced exactly this problem six hundred years ago. Towards the end of the fourteenth century, the nobility shifted from a form of partible inheritance (all children inherited equal shares of the family estate) to a system of primogeniture (oldest sons inherited everything). The main reason for this was that they were running out of new land to acquire. Partible inheritance leads inexorably to poverty within just a few generations if family estates are repeatedly broken up without the possibility of acquiring new land. So, rather than destroy their economic power, the landed families gradually preferred to invest everything in one son.

But within just a few generations, Portugal began to have problems from growing numbers of disgruntled younger sons of the nobility who were unable to find brides because they lacked sufficient resources to be attractive as prospective husbands (and strict social rules prevented them from marrying into the 'lower' classes). Roving bands of disaffected upper-class youth began to play havoc with civil order. In the end, the crown had to intervene. Their preferred solution was to encourage these young Turks to seek their fortunes abroad – in the wake of Columbus, Vasco da Gama, and Magellan's first circumnavigation of the world. In doing so, they precipitated the great age of European exploration. The burial records of the Portuguese nobility from this period bear stark testimony to this: oldest sons typically died on their estates in Portugal, but as the fifteenth century gave way to the sixteenth, younger sons died increasingly in Africa and further afield.

If human populations are left to their own biological devices, things will probably turn out all right in the long run. One of the fundamental laws of Darwinian evolutionary theory tells us that populations will usually value the rarer sex more – which is why, over the long term, sex ratios tend to approximate 50:50. A population sex ratio that gets out of kilter will, in due course, be brought back into line because parents will eventually favour the rarer sex. The problem for the Chinese, of course, is that this is something that can only happen on a timescale of many generations, even millennia. The social crisis they face requires a solution on the scale of decades at best.

The Chinese government has not been slow to appreciate this. They have been running a vigorous campaign to persuade citizens that 'girls are good too'. They have recently also threatened severe punishments for clinics that tell parents the sex of their unborn baby when the mother comes in for scans or tests. But these are long-term solutions that will take a generation or more to balance the books. In the meantime, China may have much more serious social problems to deal with. If we think we have a problem with young males and a gang culture here in the UK, spare a thought for China a decade or two hence when that problem is exacerbated by the addition of forty million sexually disgruntled young men – and there isn't the benefit of an empire to off-load them to . . . Or is economic migration to the West their answer?

Chapter 10

The Darwin Wars

It is a century and a half since the publication of Darwin's seminal book *On the Origin of Species*, yet the debates about evolution and Darwinism continue to be as lively as they were the day after the book was published. It is still very much head-to-head between science and religion, although it has to be said that it is largely fundamentalist forms of the Abrahamic religions which seem to be especially troubled by evolution. Nowhere has this tussle of world views been so publicly debated as in the USA. It was much to the joy of the evangelical Christians that, in the penultimate year of his reign (that sounds almost biblical, doesn't it?), President Bush placed his weight behind a proposal to include the theory of Intelligent Design in the American school biology curriculum.

How intelligent is design?

So what's all the fuss about? Well, many would regard Intelligent Design as creationism by the back door. It looks suspiciously as though the US educational system is turning the clock back nearly a hundred years to one of the most bizarre trials in American legal history – the prosecution in 1926 of schoolteacher John Scopes by the State of Tennessee for teaching the theory of evolution, in contravention of a newly enacted state law.

Intelligent Design (or ID) argues that the natural world is so complex that it could only have come into being if some unseen intelligence had designed it that way. In contrast, the theory of evolution – which of course eschews any such suggestion – is seen as inadequate, and full of intellectual as well as factual holes. ID is not, in fact, an especially novel idea: it dates back to the English theologian William Paley, who, in his classic 1802 book *Natural Theology*, used the perfection of nature as an argument for the existence of God (the 'grand designer').

In the words of one of ID's leading lights, the biochemist Michael Behe from Lehigh University in Bethlehem, Pennsylvania (that's almost a biblical giveaway, isn't it . . .), something as complex as a living cell could not have evolved by a series of small steps in which its elements were gradually assembled one by one: a cell without its organelles, for example, would be about as functional as a mousetrap before the spring was added. The gauntlet thrown down to the evolutionists is to show that a blind process of mutations could produce the kind of complexity we see in the world around us. Failure to do so is taken as implicit support for the default position (i.e. there must have been a designer).

To the naïve, these arguments sound extremely plausible. But their plausibility rests on a deliberate sleight of hand. Take the eye, for example. Could one imagine an imperfect eye that lacked a lens? How could such an eye possibly help its owner? Well, the short answer is that there are, in fact, plenty of examples of eyes of this kind in nature, and they are all perfectly functional and no doubt much appreciated by their owners. Eyes have been independently 'invented' at least a dozen times in different groups of animals: as a result, they take myriad different forms. We need look no further than the humble mollusc to see eyes that range from simple light-sensitive clusters of cells, to lens-less eyes, to eyes with fixed lenses, to eyes with adjustable lenses hardly different from our own.

The problem is that most of the advocates of ID seem not to be particularly well versed in good old-fashioned natural history. As a result, they are not familiar with the many everyday examples that make nonsense of their arguments. Nor, it seems, are they especially well versed in what the theory of evolution actually says. A common belief among IDers is that Darwinian evolutionary theory assumes that the process of evolution is a consequence of blind chance – mutations randomly producing small changes that gradually add together. Hence, the common claim that evolution by natural selection is equivalent to asserting that a whirlwind could assemble a jumbo jet by blowing through a junkyard. Alas, evolution is not a random process in this sense. Mutations certainly occur at random, but the processes that select and gradually fit mutations together over time are far from random: natural selection, Darwin's great contribution, is a *very* directed process and can work with astonishing speed. It has taken only ten thousand years to produce the snow-white polar bear from its common ancestor with other Eurasian brown bears.

What makes all this so intriguing is why otherwise perfectly rational people with solid scientific credentials should be so enamoured of ID. It is conspicuous that most of those who do espouse ID are not organismic biologists. For the most part, they are working in disciplines whose activities are largely unaffected by whether or not the theory of evolution is true. So why are they *so* antagonistic to Darwin's theory of evolution, given that this is in fact the second most successful theory in the history of science – after quantum mechanics in physics, which, unlike Darwin's elegantly simple theory of evolution, is perhaps the most inscrutable theory ever invented by the human mind?

We could write all this off as a kind of idle senior-common-room chitchat among those with too much time on their hands. But failing to understand the force of natural selection and its role in evolution has had, and will continue to have, rather serious consequences for all of us. It has been failure to understand evolutionary processes that gave us DDT-resistant insect pests in the 1950s, drug-resistant malaria in the 1980s, and most recently of all the terrifying phenomenon of the MRSA superbug. We really don't want any more of these than we can help.

The evolution wars

In most cases, of course, the culprit is religious fundamentalism: a desire to believe that the story of creation as set out in the Bible is literally true. But why is it that some religions have such a hard time with the theory of evolution? Why should the fact that humans have an evolutionary history that stretches back to a common ancestry with the apes exercise so many of them so much? Recently, it was the bishops of Kenya (or at least some of them) who got hot under the clerical collar. These worthies objected to a new display of our fossil ancestors' bones in the National Museum in Nairobi lest the sight of them contaminate the minds of visiting children. Bishop Boniface Adoyo and his evangelical friends feared that the poor naïfs might actually come away thinking that we are descended – Heaven forfend! – from apes!

Ever since the celebrated slanging match in Oxford in 1860 between 'Soapy Sam' Wilberforce, the Bishop of Oxford, and Thomas 'Darwin's Bulldog' Huxley, evolution has had an unusually hard time. Creationism has never quite gone away. Indeed, in some parts of the New World it is still in particularly rude health. It is not, of course, a condition confined to Christianity. Islam has some difficulty with the idea of evolution too: since it doesn't appear in the Quran, asserting its truth challenges God's omniscience, and that's considered blasphemous.

Knowledge may be power, but the suppression of knowledge is far more dangerous. It is something we can ill afford – unless, of course, we are willing to return to a strictly peasant economy and reduce the world's current population a few thousand-fold more or less overnight. In my view, we do that at our peril. There are just too many examples where attempts to control science have had disastrous consequences and derailed national development.

The most famous is the sad history of Russian biology. In 1917, when the Bolsheviks came to power, Russian genetics was at least a decade ahead of anyone else's in Europe or America. But the Russian Marxists were suspicious of genetics: Marx himself notwithstanding, they interpreted the nascent theory of (genetic) evolution as undermining the possibility that society could be changed by education and economics – the underpinning justification for the Marxist revolution. Professors of genetics were made to sit behind empty desks, and Russian biology was handed over to one Trofim Lysenko who believed that plants could be adapted to new environments merely by stressing them. The result was spectacular crop failures and serious starvation among the peasants. Meanwhile, western genetics did not catch up with where the Russians had been in 1917 until the 1930s, and then of course they just sped ahead.

A less familiar case is the history of Islamic science. As Europe laboured under the Dark Ages, science was alive and flourishing in the cities of the Islamic empire from Andalucia in Spain to Iran far to the east. Not only did these scholars preserve for us the writings of the ancient Greek philosophers (we would know nothing of Aristotle and Plato had it not been for them), but they built on these foundations to develop modern science.

The list of their achievements is staggering. They invented algebra. The word itself comes from the second word of the title of a book by the mathematician Abu Jafar Muhammed ibn Musa: his *Hisab al-Jebr w'al-Muqabala* (literally 'Calculation by Restoration and Reduction') was published in AD 825. Meanwhile, the much maligned and completely misunderstood alchemists were laying the foundations of modern chemistry, and developing the experimental method to very sophisticated levels.

In his *Kitab al-Manazir* ('Book of Optics'), the eleventh-century scholar Hasan ibn al-Haytham developed a new mathematical and experimental approach to the study of vision and light. It was the most important book on the topic until Newton published his *Optics* seven hundred years later. Towards the end of the thirteenth century – and long, long before Newton ever set foot in his local primary school – Kamal al-Din al-Farisi demonstrated for the first time that a rainbow consists of two refractions and a reflection of light within a water droplet. And when Copernicus, the founding father of modern astronomy, calculated his planetary motions in 1515, he did so using a 'Tusi couple' invented by the thirteenth-century Persian astronomer Nasir al-Din Tusi – who just happened to be al-Farisi's tutor.

But all this came to a grinding halt in the fourteenth century, when religious fundamentalists persuaded the political powers of the day to suppress science and philosophy throughout the Islamic empire because these new discoveries challenged God's omniscience. Islamic science never recovered, and the baton was handed over instead to the monasteries of Europe whither many of the Islamic-trained scholars fled.

We just cannot afford to go down that road again.

Genetics to the rescue?

One reason why creationist arguments seem so plausible is that the fossil record is very patchy. Where, assert the critics of evolution, are the intermediate fossils that link the birds and fishes, or the primates and humans? Where, in fact, is the evidence for species gradually evolving from one form to another? It's a good question. But although palaeontologists have always had explanations for why the fossil record should be as patchy as it is (the vagaries of the fossilisation process and the imperfect sampling that it inevitably provides), such arguments look suspiciously like special pleading. However, in the past decade, dramatic developments in molecular genetics have circumvented this problem, often in quite dramatic ways.

We have suspected for some time, for example, that modern birds are in fact the surviving descendants of one small family of dinosaurs. The discovery of a number of partially feathered dinosaurs in China during the 1990s added a new sense of excitement and only served to reinforce such a view. Then in 2008 came the news that molecular genetics had confirmed that this intuition was right. Birds do belong to the dinosaur family – or should that be the other way around?

This was real-life *Jurassic Park* stuff. Chris Organ from Harvard University and his colleagues carried out the first successful extraction of DNA from a sixty-five-million-year-old fossil *Tyrannosaurus rex* – the archetypal dinosaur if ever there was one. This is no small achievement, since extracting DNA samples from fossils is a tricky business. The older the fossil, the more likely it is that all the tissue has been transformed into inert stone. And even where some usable tissue has survived, the chances that the DNA can be extracted are at best poor because DNA degrades relatively quickly with time. The chromosomes break up and what you are left with is fragments of DNA that are often too short to be matched up against the DNA strands of another species.

Even then, undertaking a genetic analysis is not straightforward. You have to find the right bits of the chromosome to do these analyses. You need sections that do not code for functional parts of the body, because functional genes are subject to rapid and dramatic change under the influence of natural selection. Instead, you need chromosome segments that have no function and so change only as a result of random mutations, remaining in place because they neither benefit nor hinder the animal in its daily life. It is these that provide the basis of the 'molecular clock': by painstakingly determining how many of the base pairs in the DNA strand have mutated in each lineage since two species last had a common ancestor, we can determine how closely related they are and, more importantly, when they last shared that common ancestor.

So, armed with samples from a North American *T. rex* and a mastodon, Organ and his colleagues compared the DNA sequences for these two giants of the past with DNA from a range of living animals, including birds (represented by the humble domestic chicken and the ostrich), some primates (humans, chimpanzees and macaque monkeys), cows and dogs, rats and mice, modern elephants and a selection of reptiles, amphibians and fishes.

The genetic evidence places the mastodon just where we expect it to be (with the elephant), which gives us some confidence in the analyses. The real gem is the fact that it places *T. rex* right alongside the two birds in the sample (the chicken and the ostrich). In fact, so close is their relationship that a sophisticated statistical analysis is unable to distinguish between the three of them. More intriguingly, it includes the alligator in this group, well separated from the other reptile in the sample (the humble lizard). Alligators, it seems, might also be dinosaurs in disguise – though, to be fair, we know that the crocodile family is very old (it overlapped in time with the dinosaurs for the better part of 150 million years).

Although anatomists have come to suspect that birds and dinosaurs share a common ancestry, this example is still a reminder that appearances can easily deceive. Just because two species look very different, it does not necessarily mean that they are unrelated. The big surprise of the 1980s was the discovery that, despite the radical difference in appearance, we humans share a recent common ancestry with the chimpanzees (and to a lesser extent the gorilla). In fact, the two subspecies of gorilla (the eastern and western) are genetically more different from each other than humans are from chimpanzees. That's a sobering thought. Previously, taxonomists had assumed, on the basis of solid anatomy, that chimpanzees, gorillas and orang utans formed one ape family and humans a separate one, with a common ancestry around eighteen million years ago. The genetic evidence revealed that, in reality, it was the orang utan that was the odd one out – it did indeed share a common ancestry eighteen million years ago with the other great apes, but that was long before the three African ape lineages (human, chimpanzee and gorilla) appeared on the evolutionary scene.

So who owns your bones?

Nothing is more contentious in the museum world than the hundreds of thousands of human skeletons that lie within their vaults. What has made these bones especially contentious is the fact that most of them come from native peoples in countries where the aboriginal inhabitants have long been oppressed into the margins of modern society. And it is not always just bones. It's barely a decade since the Glasgow Museums repatriated a 'Ghost Dance Shirt' that had been taken from the body of a Sioux Indian after what was probably one of the least savoury incidents in American history, the infamous Battle of Wounded Knee in 1890.

However, few cases have been quite as curious as that of Kennewick Man. Discovered by chance in 1996 on the bed of the Columbia River, in Washington State in the USA's northwest, this virtually complete male skeleton very quickly aroused controversy when archaeologist Jim Chatters, into whose hands the bones were consigned for analysis, declared them to be about nine thousand years old – and probably of European origin. As the oldest complete human skeleton ever found in the Americas, that was inflammatory stuff. As it happens, there is now quite compelling evidence to suggest that the earliest inhabitants of North America did in fact come from Europe (the vicinity of Spain, as it happens) sometime around twenty thousand years ago. It seems that they were swamped five thousand or so years later by the arrival of the ancestors of the modern Native Americans who came from Siberia across the Bering Strait . . . But that's another story.

Native Americans, like Australian Aboriginals, have at times been very vociferous in demanding the return of all bones for reburial, on two separate grounds. One is an understandable cultural belief that the ancestors should be treated with due respect and buried properly in the safekeeping of their descendants. Many of the Native American skeletons in US museums were, not to put too fine a point on it, removed from ancient tribal burial grounds without so much as a by-your-leave. The other is the rather more murky issue of land claims. Nowadays, showing that your tribe lived at a site in earlier times gives considerable grist to the land-rights mill, and can be very big business if prior ownership of the land then allows you to build a casino there.

Now, as it happened, the land that Kennewick Man was found on was federal land under the control of the US Army. They promptly impounded the bones, but, when presented with a request for repatriation to a consortium of local tribes, agreed to hand them over. However, a group of anthropologists sued to prevent the bones being repatriated for reburial until there had been an opportunity to study them in more detail. That was in October 1998, and the case remains unresolved. One unexpected benefit to come out of all this is that, perhaps because of all this furore and the need to figure out just who he was, Kennewick Man's bones have been studied in more excruciating detail than almost any other human remains other than genuine fossils. After all, if he really is European, Kennewick Man has rather interesting implications for the history of American colonisation.

However, the issue raises the tricky question of who has rights over human remains. In one sense, the older the bones are, the more they belong to all of us. But even the most recent historical specimens can tell us a great deal about the story of our collective history, the patterns of migration, the successes and failures of our species, the trials and tribulations of human experience through the ages. Nor is this simply a matter of a quick anatomical description, or extracting a scrap of bone to analyse its DNA. Much of what we can do depends on the questions we have learned to ask, and these become more sophisticated as our knowledge grows. As every amateur archaeologist knows, much was lost for ever by poor excavation techniques even as recently as the 1940s. Moreover, the questions of yesteryear often prove to be naïve and misleading. And much depends on the discovery of new technology: DNA analysis has revolutionised our understanding of many aspects of history in the last decade or so. But we can only learn from this if the bones are there to study.

Many have complained that much of the pressure for repatriation has come from earnest but politically motivated western intellectuals, rather than from native peoples themselves. Museums – often confused about their own role in modern society, and sometimes under pressure from governments – have been over-anxious to be seen to be doing the right thing. But the outcome has sometimes been comical. One attempt to repatriate and bury four Inuit bodies that had been marooned in a major US institution, for example, was greeted with embarrassment by the Greenland community who were forced to accept them. What have they got to do with us, they asked?

Although the battle for bones has often been seen as a conflict between western science and the sensitivities and rights of native peoples, it need not always be so polarised. When the contents of the burial vaults from Christ Church, Spitalfields, in London were removed to the Natural History Museum, researchers were able to integrate their study of the skeletons with detailed family history information – sometimes even portraits – provided by descendants who took great delight in being part of the process. If more was done to persuade the communities concerned to be part of the process of science in exploring and celebrating their own histories rather than locking those histories away from sight, we might all benefit. More importantly, it might even lead to a wider understanding of Darwin's theory of evolution.

Chapter 11

So Near, and Yet So Far

Our history is a long one, stretching back some six million years to the point at which our ancestors parted company with the other members of the African great apes, the biological family to which we humans belong. The passage from then to now, however, has been far from simple or straightforward. There were many blind alleys that led nowhere in the end, even though some of them prospered for many hundreds of thousands of years before going extinct – the many australopithecines (the apemen that diversified into more than a dozen different species between six and two million years ago), the early *Homo erectus* species that migrated out of Africa and colonised Asia as far east as modern Beijing, the iconic Neanderthals of Europe. There were, equally, many moments when the fragile lineage that eventually gave rise to us teetered on the brink of extinction. The genetic evidence now indicates that all modern humans are descended from as few as five thousand breeding women who lived around two hundred thousand years ago in Africa. So small a breeding population could easily have disappeared without trace.

In fact, we live in rather privileged times. We are the

only species of our lineage now in existence. But in fact this is the first time in our lineage's six-million-year history that this has been true. The last ten thousand years or so have been unusual in having only one species of our lineage alive: prior to that, there have always been several, sometimes as many as six. Many of these now-extinct species survived a great deal longer than we humans have done so far. More sobering is the fact that some of the now extinct members of our family survived late enough to be within handshaking distance of us. The last Neanderthals died out in Europe only twenty-eight thousand years ago. The last *erectus* hominids died out in China some time after sixty thousand years ago. And on the Indonesian island of Flores, a diminutive member of this group may have survived until as recently as twelve thousand years ago. Just who were these relatives of ours?

A little lady and her long-lost family

We will never know her name. Indeed, we will never know whether she even had a name. But when her remains were unearthed in 2004 in a cave on the Indonesian island of Flores, she caused the kind of stir that we normally associate with Hollywood film stars. She died in complete obscurity around eighteen thousand years ago, only to be catapulted into glittering fame by a chance discovery.

Soon nicknamed 'The Hobbit', she and her kind (in fact, the remains of as many as five different individuals were unearthed altogether) excited the palaeoanthropology community and sent the world's media into something of a spin amid claims that the story of human evolution would have to be rewritten.

In fact, the truth has turned out to be a little more prosaic, though just as remarkable for all that. The Hobbit was certainly distinctive enough to be given a new species name, *Homo floresiensis*, after her home island. But what made her so newsworthy was not that she was one of our direct ancestors – in fact, we probably last shared a common ancestor with her about a million and a half years ago – but the fact that her kind had survived at all for so long.

Our current understanding of human evolution, based on the fossil evidence we have, goes something like this. After the long haul of the 'apeman' phase (typified by the famous 3,300,000-year-old 'Lucy' skeleton from Ethiopia, famously named after the Beatles' song 'Lucy in the Sky with Diamonds' that happened to be playing on the excavator's tape-recorder when her bones were unearthed), our ancestors underwent a relatively rapid gearshift into a more obviously humanlike form known to scientists as *Homo erectus* (literally 'erect man') some time just short of 1.5 million years ago. Though brain size increased quite a bit from the 350cc typical of its earlier apelike species, it was still a long way off the relatively massive 1250cc that we find in modern humans. What we do find, however, is a new body shape that has the same long legs, narrow hips and barrel chest that modern humans have – features associated with a more efficient form of striding walk that was good for covering long distances in a nomadic, migratory lifestyle.

With its body newly designed for long-distance walking, *Homo erectus* set off to conquer the world, breaking out of Africa for the first time around a million years ago, and very rapidly colonising the farthest corners of mainland Asia. For a long time, nothing much of interest happened, and there was little to differentiate between the Afro-European populations and those in eastern Asia. But in the millennia that followed, the Asian populations went their own way, cut off from their African cousins.

Around half a million years ago, some of the African populations began to undergo rapid change, mainly involving a dramatic increase in brain size and another exodus out of Africa into Europe. Then over the space of the next couple of hundred thousand years, the African populations of this new species metamorphosed into modern humans, and exploded out of Africa once again (about seventy thousand years ago). In the next ten thousand years, this new species colonised every corner of the ice-free Old World (including Australia), finally even launching itself across the Bering Strait into the Americas around sixteen thousand years ago.

When these newly minted modern humans reached the Far East, it has always seemed likely that they came into contact with the remnants of the east Asian *erectus* populations surviving in the backwaters of China long after their African equivalents had died out or evolved into the modern human form. So far as we knew, none of these Asian *erectus* populations had survived past sixty thousand years ago – just about the time that modern humans turned up on their doorstep. Given our historical record when colonising new lands, was that a coincidence, I wonder . . .?

The discovery of the little lady of Flores changed all that. There she and her kinfolk were, hale and hearty, perhaps as recently as twelve thousand years ago – a mere handshake away in geological time. Modern humans must surely have come across them in the forests of Indonesia on their way to Australia (given that they had made it over to Australia by about forty thousand years ago).

But the Hobbit and her kind were nothing if not distinctive: she was tiny. We are familiar enough today with diminutive humans – today's Pygmy peoples of central Africa and the negrito peoples of the south Asian forests are not much taller than she was. But whereas all these modern human pygmies have brains that are the same size as ours, the Hobbit and her kind had brains that were no bigger than those of our mutual apeman ancestors.

What did surprise everyone was the fact that their bones had been found beside stone tools of a modestly sophisticated kind, as well as evidence for fire and the hunting of large animals (including the now extinct, formidable stegadon and the very much still living giant lizard, the Komodo dragon). For something the size of a five-year-old human child, killing a thousand-kilogram stegadon is no small feat; at best, it seems to suggest some degree of co-ordinated planning and co-operation. Of course, it is always possible that the tools were actually made by modern humans. But if that is so, it raises the question of how the tools and Ms Hobbit and her friends got to be in the same place at the same time. The usual conclusion drawn in such cases is that the tool-makers ate the individuals whose bones we find among the tools. That's not beyond the bounds of possibility – after all, chimpanzees and gorillas are eaten with culinary enthusiasm in western Africa today, and monkeys are a delicacy in Indo-China. The Hobbits would have seemed neither more nor less than another ape to our ancestors. However, as yet, there is no incontrovertible evidence for their having been eaten – something that would normally be signalled by cut marks on bones, broken marrow bones and perhaps evidence of cooking (for example, scorch marks on the bones). So the jury is out on this one.

There is, however, one last curiosity worth mentioning. On the nearby island of Borneo, one of the largest of the Indonesian chain of which Flores is a part, the local people have long claimed that they were familiar with three kinds of people in the forest – *orang rimba* (a tribe of perfectly respectable forest people also known as the Suku Anak Dalam, meaning 'children of the inner forest'), the *orang utan* (the familiar Asian great ape) and the *orang pendek* (a diminutive forest dweller that was half man, half ape). Perhaps the orang pendek is a surviving folk memory of contacts with the Hobbit. We really must have come within just a whisker of shaking her hand.

To be, or not to be, an ancestor

Until very recently, the geological strata of Africa (or indeed anywhere else) have stubbornly refused to yield up any hominid fossils more than 4.5 million years old. However, in 2000, a French team unearthed fragments of a hominid-like creature from deposits in the Tugen Hills just above Lake Baringo in central Kenya that were dated to around six million years ago. In all, twelve fragments (including parts of limb bones, jaws, a hand bone and some teeth) representing at least five different individuals have been recovered from four sites. The specimens were named *Orrorin tugenensis* (*orrorin* means 'original man' in the local Tugen dialect), but the inevitable nickname

'Millennium Man' soon caught on.

The following year and a thousand miles to the west, another French team, which had been fossil-hunting in west Africa for the better part of two frustrating decades, turned up a near-complete skull and some jaw and teeth fragments at a remote site on the southern edge of the Sahara in Chad. It had a slightly older date (between six and seven million years old). Nicknamed *toumaï* (a name for a child born dangerously close to the dry season in the local native dialect), the species was formally named *Sahelanthropus tchadensis* (literally, the 'Sahara ape-man from Chad').

Dates of around six million years place both finds within the timescale for the common ancestor of modern humans and chimpanzees suggested by the molecular data. This was getting pretty exciting.

The *orrorin* material includes two well-preserved partial femurs (thigh bones), which are similar in shape to (but considerably larger than) the diminutive femurs of the earliest australopithecines. Although claimed as evidence of bipedalism, it is difficult to be certain that these femurs really are from a bipedal walker rather than a more conventional quadrupedal great ape because the lower ends are missing. In modern humans (and all uncontroversial hominids), the shaft of the thigh bone is angled outwards when the knee joint is placed on a flat surface. This allows the body's centre of gravity to lie directly above the foot that is in contact with the ground at any given moment when striding. In contrast, the shafts of all habitually quadrupedal living great ape femurs are vertical – which causes them to waddle awkwardly when walking on two legs.

Although the leg bones do not rule out bipedalism, the fragment of an upper arm bone from Tugen shows some similarities to those of living chimpanzees and suggests a partially arboreal lifestyle. The suggestion of an arboreal life is reinforced by the curved shape of the finger bone, a feature that is characteristic of tree-climbing great apes but not modern humans.

The slightly older Chad material has been the subject of much more controversy. The discoverers claimed that the species is the oldest known member of our lineage on the grounds that the remarkably complete skull shows features (brow ridges and small canines) that are only found in early members of the genus *Homo* (which date from around three to four million years later). Although the front of the face does share some resemblances with those of later hominids, the skull looks like most other ape skulls when seen from behind and its cranial volume (approximately 350cc) is well within the range for modern chimpanzees. More importantly, the foramen magnum (the hole in the base of the skull through which the spinal cord passes on its way from the spinal column into the brain) seems to be positioned towards the back of the skull (as in living great apes) rather than in the centre of the skull (as in living humans – and all known fossil hominids whose skulls are balanced on top of a vertical spine). This rather suggests a quadrupedal style of locomotion more like that of living apes.

Despite these uncertainties, it is clear that both *toumaï* and *orrorin* represent important members of the African great ape family at the critical juncture when the hominid lineage was parting company with the chimpanzee lineage. One aspect of their biology is of particular interest given that at some early stage our first ancestors moved out of the forests still preferred by all living great apes and moved into more open, wooded habitats: the presence of antelope and colobine monkey fossils at the same site as *orrorin* is indicative of a wooded rather than a forested habitat, suggesting that a number of these early ape species may have been venturing into this new world.

These two new fossils point to two key conclusions. First, there seem to have been several different species present at around the time of the hominid–ape split. And, second, these various species were very widely distributed – and living in areas like central Chad that are now far from the forests occupied by contemporary great apes, the nearest of whom live some four hundred miles (650 km) to the south.

Visions in stone

Meanwhile, back in Europe, we were missing another opportunity to shake hands with our past – this time, in the form of the artists who created the magical prehistoric cave paintings of Spain and southern France.

Our story begins one day in 1879 when a bored young girl out exploring a cave with her father, the local landowner Don Marcelino Sanz de Sautuola, chanced to look up at the ceiling. She made a spectacular discovery. Above her, bison, deer and horses turned and twisted, bunched up against each other fighting for space, or lay chewing the cud, just as they had been left eighteen thousand years before by the prehistoric painters who had made them. This cave, at Altamira in northern Spain, has turned out to be far from unique: there are around 150 known sites of prehistoric cave art in Europe. And the artwork is little short of exquisite. It is easy, in the dark of these caves, to become lost in the mystery of the figures that some unseen hand sketched so long ago. Grown men have been reduced to tears before them.

Here, in one corner of an ancient gallery, is a child's hand, stencilled around by paint blown from the mouth. If the guardians of the cave would allow it, you could place your own hand over the outline, and reach out across the millennia to touch that child. A delicate, hesitant touch, such as one might give to a new lover. It is impossible not to feel the magic in the air. Who was this child? By what name were they known? And what became of them? Did he or she grow up, have children of their own, and live to a ripe old age, a respected white-haired member of the community, remembering in the misty twilight years of old age a day – one spring, perhaps – when they had been led down the winding tunnels by the dim light of a tallow lamp to a remote back chamber and made to press their hand against the cold wall of the cave while one of the men blew paint across it. Or, instead perhaps, did they die of some childhood illness or accident, or fall prey to a wandering predator – a future cut off in the first flush of childhood, one of many small tragedies in the life of its mother, each marked by the anguish of loss, its passing signalled by a shrill brittle halo of inconsolable wailing.

We shall never know. But what we can say is that the people who made these drawings engaged in life with an exuberance that resonates with us today. Cave art is the final flowering of a remarkable development in human evolutionary history, a phenomenon that archaeologists refer to as the Upper Palaeolithic Revolution. It began around fifty thousand years ago with a sudden burst of very much more sophisticated stone, bone and wooden tools – including needles, awls, fishhooks, arrow- and spearheads.

From around thirty thousand years ago, this is followed up by a veritable explosion of artwork that has no particular function in terms of everyday survival but seems rather to be entirely decorative. There are brooches, carved buttons, dolls, toy animals and, most spectacular of all perhaps, figurines – exemplified above all by the so-called Venus figures of central and southern Europe. These famous 'Michelin-tyre' ladies seem to have been the pin-ups of their day. Big-hipped and ample-bosomed, with their hair often beautifully braided, these ivory and stone (sometimes even baked clay) statuettes are quite the most spectacular of the late Palaeolithic artefacts.

Then, from about twenty thousand years ago, we begin to find evidence for deliberate burials, for music and for a life in the mind. The cave paintings of Altamira, Lascaux, Chauvet and the many other grottoes, shelters and caverns across southern Europe and beyond are but the icing on this grand artistic cake. Nothing like it had ever been seen in the history of human evolution. Buried within it lay the foundations for modern human culture, from literature to religion and, beyond, to science.

This outpouring of craftsmanship speaks to us across the intervening millennia. Here are a people who are not so very different from ourselves: what we find beautiful, they too found beautiful. Here, it seems, encapsulated in a brief moment in time is the essence of what made us who we are, what finally produced humans as we know them, with all that inflorescence of culture that makes us in some intangible but very certain way utterly different from every other species alive today – and, indeed, every other species that preceded us in the long history of life on earth.

The mysterious Neanderthals

When the ancestors of the Altamira cave artists arrived in Europe some forty thousand years ago, they did not find an empty continent. Europe had already been home to the Neanderthals for two hundred thousand years. The Neanderthals were an exceptionally successful race of humans whose ancestors probably arrived in Europe around five hundred thousand years ago. Over the following few hundred thousand years, they gradually developed the characteristic Neanderthal form – a thickset, very heavily muscled body, a large head with its characteristic 'Neanderthal bun' (or bulge) at the back, a heavy chinless jaw and massive nose. In this form, they successfully colonised the plains of Europe as far east as the Urals. There, they hunted large game (including the much fabled mammoths) by the very risky strategy of impaling their victims on heavy thrusting spears. Not for them the lightweight, javelin-like hurling spear or the bow and arrow later to be favoured by our own immediate ancestors.

When the last Neanderthal died (probably in northern Spain) less than a thousand generations ago, they had been around as a species for a great deal longer than we modern humans have so far managed. Modern humans emerged about two hundred thousand years ago from the same African root stock as the Neanderthals. But unlike the Neanderthals, we remained in Africa until around seventy thousand years ago when there was a sudden exodus across the Red Sea into southern Asia. Modern humans did not reach Europe, where they came into contact with significant Neanderthal populations for the first time, until around forty thousand years ago. When they finally did so, they arrived – as have so many of Europe's historical immigrants from the Indo-Europeans around six thousand years ago to Attila the Hun and his nomad hordes in Roman times – from the steppes of western Asia. It took us little more than ten thousand years to displace all the Neanderthals from Europe.

The sudden disappearance of the Neanderthals has always piqued our curiosity. Some have suggested that they disappeared because modern humans bred with them – modern Europeans being thus the result of hybridisation between the two species. It's true that very occasionally you do get the odd Neanderthal-like modern European, complete with barrel-chest, thick neck and heavily muscled legs and arms. But that said, there are too many thin gangly ones who don't show much of a resemblance, and on the whole this seems a rather implausible explanation. Others have suggested that, on the model of the historical European invasions of the New World and Australia, our ancestors simply slaughtered the Neanderthals because they were in the way or put up a resistance. Alas, we modern humans have rather a bad history of such behaviour, so it's by no means beyond the bounds of possibility. Others have suggested, in the light of the more recent experience of the South American Indians, that the Neanderthals were wiped out by novel tropical diseases brought from Africa to which they lacked immunity. The only fly in this particular ointment is that modern humans didn't arrive directly from Africa: they came from the east, probably somewhere around the Black Sea, so had been exposed for the better part of thirty thousand years to much the same diseases as the Neanderthals would have had.

Whatever the actual cause of their demise, the Neanderthals probably viewed these darker-skinned immigrants with much the same suspicion that modern Europeans have done in more recent times. That the Neanderthals were light-skinned like modern Europeans received dramatic confirmation from recently published analyses of Neanderthal DNA. Geneticists at the University of Barcelona have managed to extract DNA from a forty-eight-thousand-year-old Neanderthal from El Sidrón in Spain. There they found a variant of the MC1R gene that, in modern Europeans, is responsible for lighter skin colour by suppressing the production of dark melanins in the skin. When copies of this gene are inherited from both parents, the result is the sun-sensitive skins and red hair that have been such a hallmark of our own west-coast island populations. Red-head Neanderthals? That's a turn-up for the books.

At the same time, it is equally clear from these and other recent genetic studies that Neanderthals shared few of the novel mutations that characterise modern human populations, especially those from the northern hemisphere. It seems that the Neanderthals were not our ancestors, but a separate – albeit closely related – species. Our European light skins and red hair were not the result of our dark-skinned African ancestors interbreeding with Neanderthals, but rather independent genetic adaptations to coping with the same problems of life at high latitudes that bedevilled Neanderthals. It's that old vitamin D problem that we came across earlier.

One reason this must be true is that the genetic evidence has now comprehensively confirmed that the Neanderthals' ancestors split away from the ancestral lineage that gave rise to us around 750,000 years ago, some time before the lineage that eventually gave rise to the Neanderthals first left Africa in search of a new homeland in Europe. Whatever the root cause of the Neanderthals' sudden demise some four hundred millennia after arriving in Europe, the one that can now definitively be ruled out is interbreeding with modern humans. The alternatives that remain, however, would seem to be a lot less pleasant to contemplate.

Chapter 12

Farewell, Cousins

Species change through the gradual failure of some lineages to reproduce, resulting in a subtle but steady drift in the species' genetic make-up towards that of lineages that are more successful. Although in most cases these processes are quite slow, an entire species can go extinct catastrophically if none of its various lineages can reproduce fast enough to offset unusually high levels of mortality. There is always a steady trickle of such extinctions over time – there have been literally dozens within our own lineage during the course of our six-million-year evolutionary history. Sometimes, however, environmental conditions conspire to produce a rapid burst of extinctions.

Farewell, cousins . . .

Sixty-five million years ago, a massive asteroid smashed into the corner of Mexico where the Yucatan peninsula now stands. The resulting fireball, combined with millions of tons of vaporised rock thrown up into the atmosphere, brought on a nuclear winter that changed the face of the earth for ever. As the planet slowly emerged from the catastrophe, it was to find that the dinosaurs who had dominated the planet for the previous 250 million years were fading fast. The dragon lords of the earth were being replaced by a small and insignificant group of animals – the mammals – that had previously skulked out of sight on the forest floor.

This dramatic turnover in the world's fauna was the fifth massive bout of extinction in the five-hundred-million-year history of life on earth. Most of these mass extinctions seem to have occurred at intervals of about sixty-five million years. Although their causes seem to have varied, they have typically resulted in the sudden disappearance of seventy to eighty per cent of all the species of animals alive at the time.

So it is, perhaps, no surprise to find ourselves on the brink of yet another wave of extinctions. Although only a relatively small number of species have actually gone extinct in historical times, many are famous for having done so – the dodo of Mauritius and the giant moas of New Zealand are the best known, but the curiously named Miss Waldron's red colobus from the Gambia and the giant lemurs of Madagascar (some as big as a female gorilla) remind us that even primates are not exempt.

But the figures for actual extinctions give a false impression. Some eleven thousand species of animals and plants are currently listed as being in imminent danger of extinction. The latest estimate is that as many as half of all living species could be extinct within the next century. Sadly, the cause this time is not meteors from outer space or poisoning from volcanic eruptions from within, but – to borrow the Gaelic for a moment – *sinn féin*: we ourselves.

We have been cutting down the world's forests at such a rate over the past century that some African countries now have as little as five to ten per cent of their original forest cover left. What remains of the planet's forests are being lost as a rate of about eight per cent per decade. It hardly needs rocket science to figure out what that means: it won't take much more than a century to clean up the rest.

The tragedy that hides beneath these bald figures is brought sharply into focus by the prospects for our closest living relatives, the great apes. If you want to see an orang utan in the wild, you'd best book your plane ticket now. The rate of deforestation in their strongholds on Sumatra and Borneo, and the resulting decline in orang numbers, is such that there are unlikely to be any left in the wild in 2015. And the Boxing Day tsunami didn't help, either: the Aceh peninsula in northern Sumatra, which took so much of the brunt of the human tragedy, was also one of the strongholds for wild orangs. Even before the tsunami struck, the peninsula was estimated to have lost forty-five per cent of its orang population between 1993 and 2000 alone.

The forecasts aren't much better for their African cousins. The gorilla and the chimpanzee, with whom we shared a common ancestor as recently as six or seven million years ago, will outlive their Asian cousin only by a few decades. A lethal combination of deforestation and hunting to feed a voracious market for 'bushmeat' in the cities of central and west Africa holds out a promise of only another twenty to fifty years for most wild populations.

The root cause, in the end, is the dramatic explosion in the human population over the last two thousand years. When Jesus Christ was born, the world's entire population amounted to about two hundred million people (less than the current population of the USA); today, there are over 6,400 million of us, and we are adding around seventy-four million new ones every year – a baby every three seconds. Most are living in such grinding poverty that they cannot afford the luxury of worrying about conservation. The tree that is still standing quite literally stands between them and daily survival: cut down, it represents money, fuel, food or housing.

Like the proverbial car crash in slow motion, we stand on the edge and watch a disaster unfolding with an apparent inevitability that is difficult to comprehend, let alone do anything about. With or without the Kyoto Agreement, we have to sort out both our insatiable appetite for hardwoods and the demand for new agricultural land. Here, on a global scale, is the same crisis of survival that triggered the emigrations and Clearances from the Highlands and islands of Scotland during the late eighteenth and early nineteenth centuries. But in the 1800s, the emigrants had somewhere else to go to begin a new life. Today, we don't have that luxury.

Frankincense on hold

Remember the Three Wise Men, and gifts they brought the first Christmas? Gold, frankincense and myrrh? Alas, it seems that had it been this year rather than two thousand or so years ago that they popped down to the local market for a few things to take with them on the way to Bethlehem, one of the boxes carried onto the stage at primary-school nativity plays today by three puzzled waifs might have been very different. We are busily killing off the tree that produces the sticky sap that, once dried, we call frankincense. And this has bigger implications than merely for the Magi and a few school nativity plays. Frankincense remains one of the core ingredients for the perfume industry, as well as in its more conventional use as incense. Frankincense production is falling: the sap is becoming harder to get hold of.

Frankincense is produced by a handful of tree species of a small and rather undistinguished kind that grow in the arid zone bordering the southern edge of the Sahara. Like many tropical trees, the *Boswellia* exudes a sticky sap when it is cut or damaged. The sap helps protect the tree against desiccation, bacterial and fungal infections and insect predators while the damage is repaired. However, *Boswellia* sap has some unusual properties which set it aside from most other species. The dried sap yields a headily aromatic fragrance that is prized for its perfuming capacities. It was not long before people discovered that sap production could be encouraged by deliberately cutting the tree bark. The sap that oozed out could be collected a few weeks later, and this cycle could be repeated over and over again.

French crusaders were probably responsible for bringing it back to Europe from the Holy Land during the Middle Ages – hence its name, the Franks' (or French) incense. But it had been used in the Middle East for millennia as a ceremonial and general household incense, as well as in traditional herbal medicines. Incense has been a major industry throughout the tree's natural range, but especially so in the Horn of Africa and Arabia, probably for as long as humans could light the fire to burn it on.

Alas, there's no such thing as a free lunch in real life,

and especially so in the biological world. Sap production is a hugely beneficial capacity for a tree, since it provides protection for damaged parts and so aids recovery and regeneration. But producing the sap is actually very expensive for the tree. It has to take energy and resources away from reproduction the following season in order to do so. Sap, fruits and flowers all have heavy carbohydrate bases, so if the tree is forced to invest its limited carbohydrate reserves in sap, it simply doesn't have these available for investing in flowers and fruits when the production season comes around with the following rains. The cost to the tree is especially heavy if its sap is harvested during the dry season: it has to draw on its stored reserves of carbohydrates to produce sap since it cannot create new carbohydrates through the natural processes when it is dormant.

In a recent study, Toon Rijkers and his colleagues at Wageningen University, Holland, and the University of Asmara, Eritrea, looked at the regeneration of the frankincense-producing *Boswellia* trees in the Horn of Africa. They found that the more heavily trees are tapped – and in the most intensive harvesting, the wounds are reopened every three weeks throughout the long dry season – the poorer was the flower and seed crop produced the following wet season.

Rijkers and his colleagues also found that heavily harvested trees produced seeds that weighed much less than those produced by less heavily harvested trees. More importantly, these smaller seeds had much poorer germination rates. In experimental tests, fewer than forty per cent of the seeds of heavily harvested trees produced viable seedlings compared to around ninety per cent for trees that had not been harvested for more than a decade.

In short, the demand for frankincense has been literally bleeding the trees to death. Unable to seed properly, they have not been able to replace themselves as natural mortality has taken its toll on the adult trees. However, all need not be lost: Rijkers showed that providing the harvesting is done more sensitively and the trees allowed to rest from time to time, they will regenerate well.

Unfortunately, as with all sustainable harvesting schemes, the pressures of economics and everyday survival hover ominously in the background. In the poorer countries of the world where life is on the margin, over-exploitation of one's natural resources is always a temptation. The problem that confronts most of the people is simply making it through to tomorrow: the future will just have to look after itself. If destroying a natural resource like the *Boswellia* trees allows you to survive today, that's better than starving while you admire a healthy stand of trees. This natural human instinct is the central problem in conservation. Until we can all enjoy a reasonable standard of living everywhere, the planet will always be fighting a losing battle against the forces of day-to-day survival.

Who did for the mammoths?

If there is one iconic picture of Ice Age humans, it must surely be that of half a dozen muscled prehistoric cave-men surrounding an angry mammoth which they are trying to spear to death. In the background, there is always a herd of these monsters ambling away into the distance across the tundra, apparently unconcerned. And so it may have been. But the sad reality is that these uniquely northern-hemisphere members of the elephant family (they occurred in North America as well as Eurasia) eventually went extinct. Mind you, it is sobering to remember that mammoths were still living on Wrangel Island in the Siberian Arctic just 3,700 years ago.

The classic explanation for the demise of the mammoths was that they were hunted to extinction by humans invading the tundras of the north in the wake of the retreating Ice Age – a phenomenon sometimes known as the 'Pleistocene Overkill'. The main evidence was that many large animals, including mammoths, disappeared from North America shortly after the first Native Americans arrived some sixteen thousand years ago. But a more recent suggestion has been that it was climate warming that made it impossible for these lumbering giants to find enough food. It has always been difficult to decide between alternative explanations for past events of this kind. However, an answer might now finally be at hand, thanks to the wonders of modern computers. This has come about through a combination of better climate models that allow us to reconstruct past climates, and a better understanding of the mathematics of conservation biology.

David Nogués-Bravo of Madrid's National Museum of Science and his colleagues used powerful new climate models to backtrack over the last 130,000 years and reconstruct the climate over the mammoth's entire continental range in Europe and Asia. They used these to determine the climatic conditions that would have been found at all the sites where mammoths are known to have occurred. Their findings suggest a gradual increase in the size of the area with climates suitable for mammoths between

127,000 years ago and forty-two thousand years ago, followed by a long period of climatic stability during which the mammoth's geographical range extended down into southern China and even into modern-day Iran and Afghanistan. But between twenty thousand and six thousand years ago, the climate warmed precipitately and by six thousand years ago the mammoths would have been confined to the rim of the Siberian Arctic and a few isolated places in Central Asia.

This marked reduction in mammoth-friendly habitat would inevitably have coincided with a dramatic collapse in the size of the mammoth population. And it is at this point that humans become important. Modern humans had been hunting mammoths ever since they first came across them after breaking out of Africa for the first time some seventy thousand years ago. Nogués-Bravo and his colleagues used mathematical models from conservation biology to estimate the mammoth's susceptibility to hunting pressure under different kill regimes and population densities. During the phase when mammoths were most abundant, between forty thousand and twenty thousand years ago, human hunters would have had to kill in excess of one mammoth per person in the population every eighteen months to drive the mammoth populations to extinction. But during the later phases around six thousand years ago when mammoth populations were at their lowest ebb, it would only have taken kill rates of less than one mammoth per person every two hundred years to wipe the species out. This is clearly so low that even very occasional hunting would have been enough to tip the mammoths over the brink.

We know from the archaeological evidence that hunting rates must have been high, for early humans living in the Ukraine fifteen thousand to twenty thousand years ago made extensive use of mammoth bones for building shelters. In some cases, bones were simply used to weigh down the edges of tents. But at Mezhirich in what is now Ukraine, four huts were built with walls and roof supports made out of the leg bones, lower jaws, skulls and tusks of many mammoths. Just these four huts are estimated to contain bones from as many as ninety-five different individuals.

So the lesson for us today is that while the mammoths could easily absorb the hunting pressure put on them by humans when they were abundant, their ability to do so changed abruptly once climate change caused a dramatic drop in their numbers. At that point, even very modest hunting pressure was enough to tip them over the edge of extinction. It remains an object lesson for us today, with the renewed threat of further climate warming putting increasing numbers of rare species at risk.

Gaelically speaking

Languages go extinct just like animals and plants, and we are currently witnessing a major period of language extinctions. Although there are thought to be just under seven thousand languages currently spoken in the world, no fewer than 550 of these are spoken by fewer than a hundred (mostly rather elderly) people and will certainly be extinct within the next decade or two. Perhaps as many as half of all the rest will be extinct within the next century. One of those could well be Gaelic, the language of the Scottish Highlands and islands for at least the past thousand years since the western seaboard was colonised by Gaels from Ireland. With only around sixty thousand, invariably bilingual speakers in Britain (ironically, there are more native Gaelic speakers in Canada, whither many Scots emigrated in the nineteenth century), Gaelic is already on the critical list: it will not take many generations of declining use in everyday contexts for it to slip over the edge of oblivion to join Latin, Sanskrit, Pictish (the language previously spoken in the Highlands when the Romans arrived in Britain) and the dinosaurs.

Should we worry?

The short answer is yes, and for several different reasons. One is the more general one that we can learn a great deal about the history of language evolution and the historical migrations of peoples from their languages. Some of the world's more obscure languages have a great deal to tell us, especially when we contrast what the language has to say with what its speakers' genes tell us about their physical movements. The two are not always the same, because languages can be acquired as a result of trade or conquest.

The history of our own European languages offers examples of every possible combination in response to conquest. The Slavic Lombards and the Germanic Franks – who invaded northern Italy and France, respectively, as the Roman empire collapsed – abandoned their native languages in favour of the more up-market nascent Italian and French of their no doubt reluctant hosts. In contrast, Attila and his Huns apparently made rather more of an impression on their hosts, who, despite solid middle European ancestry and genes, adopted with unseemly alacrity the Mongolian language of their new overlords, so giving rise to the Magyar spoken in modern Hungary. Perhaps fortunately for us in Britain, we decided to keep both our original Anglo-Saxon (a Germanic language) and the new-fangled French (a Romance language descended from Latin) brought in by William the Conqueror and his friends in 1066 – which is why English has such a rich vocabulary, since we invariably have a Saxon word (usually short and blunt) and a French word (usually long and flowery) for everything, and so can use them to create subtle shades of meaning.

Languages are also a repository for folk knowledge, some of which can be medically important (aspirin and quinine are well-known examples that were acquired from the Indians of South America). Losing a language before we have had time to search out its pearls of wisdom may lose us valuable products. Recent experiments have revealed, for example, that granny was right all along to insist on doling out chicken broth as a curative for common ailments: it turns out to be packed full of biochemically active ingredients that are very good at combating viral and other infections. Had granny's language died with her, the folk remedies figured out painstakingly over many generations by her ancestors might well have been lost for ever.

Languages also provide us with a unique window into other cultures. In this, Scots Gaelic offers an unusual example. From the great eighteenth-century poets Duncan Ban MacIntyre and Rob Donn to our own century's Sorley MacLean, an unusually rich tradition of

Gaelic poetry has graced the hearths of lairds as much as the humbler turf firesides of an evening ceilidh when the work was done. It's a remarkable tradition of oral literature, kept only partly aflame today by groups like Capercaillie and Runrig. More remarkable still, in cultural terms, are the waulking songs of the Hebrides that I mentioned in Chapter 7. No other culture has produced anything like these unique women's work songs, with their extraordinary rhythmic drive, poetic sonority, humour and sense of community. Along with the shimmering brittleness of Gaelic psalm-singing, these songs represent a remarkable cultural flowering in the Western Isles. All this will be lost if Gaelic becomes extinct.

Languages share with biological species many of the same biogeographic and evolutionary properties. Like animal species, languages are more abundant, have smaller geographic ranges and are more tightly packed near the equator than at higher latitudes. One reason for this seems to be that habitats become more seasonal and less predictable at higher latitudes, and this necessitates larger exchange networks to buffer oneself against crop failure. One consequence seems to be a form of ecological competition for niche space.

The pressure to adopt the most common language in a region (especially when backed by political force) leads inexorably to the extinction of the smaller languages. Small languages survive only where you can be self-contained and self-sufficient. For both languages and biological species, remedial action is necessary if we want to stem the extinction tide.

Extinction and the ghost of Dr Malthus

The great threat to life on earth is, and always has been, climate change. So the world breathed a sigh of relief and emerged from the Montreal climate warming summit in 2005 with at least the promise from everyone, including the USA, to take climate warming seriously and think about how we might take steps to ameliorate its worst effects. Minds were perhaps concentrated by the succession of major disasters of the previous months – the Indian Ocean tsunami, the Kashmir earthquake and Hurricane Katrina. Of the usual suite of major disasters, we missed out only on a serious volcanic eruption.

For natural disasters, 2005 was a worse than average year: some four hundred thousand people were killed by natural disasters, about five times the number killed in an average year. Still, just to set that in perspective, it's sobering to remember that well over a million people are killed each year on the world's roads, and around eight million children die of preventable childhood illnesses.

On the grander scale of the earth's history, however, dramatic changes in climate are by no means unusual. Everyone knows about the Ice Ages that intermittently engulfed most of northern Europe in massive sheets of ice. In fact, these came and went on a roughly sixty-thousand-year cycle, intermingling the grip of super-winter with rather more balmy climatic conditions. Indeed, we are in the middle of just such a balmy period now. The last Ice Age ended around ten thousand years ago with the rather dramatic Younger Dryas Event when the earth's average temperature rose by a staggering 7°C in just fifty years. It resulted in sea-level rises in excess of three hundred feet when the ice locked up in the polar ice sheets melted. By comparison, the current forecast of a four degrees' temperature rise by 2080 is quite modest.

But take an even bigger step back if you want to see just how unusual today's climate really is. Measurements of the relative abundance of different carbon isotopes in seashells indicate that between sixty million years ago (when the ill-fated dinosaurs finally died out) and around forty million years ago the average temperature of the earth was around 30°C, double its current value. Europe and North America boasted tropical forests. The earliest lemur-like primates scampered through these forests, while hippos wallowed in the steamy swamps below, right there in the heart of London, Paris and Berlin. On the longer timescale, the current cool phase is in fact quite unusual.

So whether or not our industrial and agricultural activities have caused the current warming, we would do well to remember that the earth's climate is naturally unstable. Our real problem is how we cope with these changes as they occur. The optimists will want to rely on science. After all, they might say, science has already got us out of one such mess.

Nearly two centuries ago, Thomas Malthus stirred a few feathers by pointing out that the world was heading for disaster because agricultural productivity couldn't keep step with the rate at which the population was increasing. Darwin was greatly influenced by Malthus when he was writing his *Origin of Species*: it provided him with the insight as to how natural selection might work. But not everyone was as convinced by Malthus as Darwin was. Many were sceptical and argued that the nascent sciences would solve the problem of food production for us.

As it turned out, the sceptics proved to be right, because science bought us time. A lot of frenetic activity down on the farm gave us the Aberdeen Angus and the Belted Galloway, the Blackface sheep, new and improved ploughs and seed drills. It enabled us to produce a great deal more off each acre of land than our medieval forebears could even have dreamed about, and it finally did away with the Highland 'ferm touns' and the old rig systems of medieval agriculture.

But there is a worrying difference between then and now. The agricultural revolution relied on old technology, the kind that every farmer worth his salt knew by instinct. New developments in science today depend on much more sophisticated kinds of knowledge. And the worrying point here is that the number of new discoveries per decade has been declining steadily for most of the last century. That's not too surprising: each new discovery becomes harder to win because it depends on much more complex technology, and ever deeper knowledge. The frontiers of knowledge are just becoming harder to mine, as well as being vastly more expensive.

But perhaps our real problem is that Malthus's ghost is still hovering over our shoulders. He was not wrong: it was merely that science bought us time. In the end, it is not that we are using more and more fossil fuel each decade, or carelessly dumping wastes and surpluses, but that there are just more and more of us every year wanting to do these things. It has sometimes been claimed, for example, that traditional hunter-gatherer societies were (and still largely are) natural conservationists. Unfortunately, the evidence doesn't actually support that claim. The reason traditional peoples seem to be good conservationists is simply that there are never enough of them in one place to do serious damage to their environment, no matter how badly they treat it. The rise of cities has a lot to answer for, and we would do well to learn this lesson rather faster than we have been inclined to do so far. We really do need to get the world's population growth seriously into reverse.

Chapter 13

Stone Age Psychology

Evolutionary psychologists sometimes caricature us as having 'Stone Age minds in a space-age universe'. To the extent that our minds are the product of our brains, and brains do not evolve very quickly, the ways we think and react to life's experiences inevitably reflect adaptations to circumstances long past – life as we lived it between five hundred thousand and, well, let's be generous and say ten thousand years ago when modern humans first invented agriculture and changed both their lifestyle and their environment by living in villages. The obvious implication, not lost on some evolutionary psychologists, is that we can expect much of our behaviour to be deeply out of kilter with the circumstances we find ourselves in now. In fact, maladapted, not to put too fine a gloss on it. In other words, in the vastly different circumstances of today – different because of the radical changes that culture has imposed on modern life and the environments we live in – we respond as though we were still on the plains of Africa, hunting wild game and spearing our enemies from over the hill. We respond by instinct rather than judgement. You don't believe me? Well, let me give you some examples.

The good, the bad and the tall

It is a curious fact that, out of all the job interviews I have ever had, the only two occasions on which I was actually offered the job were when I had deliberately gone out beforehand and bought a new suit. Surprising? Not at all, you might say: isn't life all about packaging? Well, of course, but we're talking about real jobs here – convincing a select group of experts is surely different from pulling the wool over the eyes of the average Joe Public.

Well, maybe not. Arnold Schumacher, of the University of Hamburg, became intrigued by the fact that successful people are often perceived as being taller than they really are. Remember how surprised you were when you finally met the queen and found out how much shorter she is than you had imagined? Schumacher put the matter to the test by measuring the heights of people at different levels of achievement.

He found that, in professions as diverse as business management, nursing and trades like carpentry, those who had achieved higher status were indeed significantly taller than those who occupied the lower rungs of the professional ladder, even when differences in age were taken into account. For example, in a sample of German business executives, senior managers were on average five centimetres taller than staff in more lowly positions, and this was true for men and women separately, irrespective of their class background and educational achievement.

Not only were successful individuals actually taller than their less successful colleagues, but success was perceived as being associated with a whole constellation of positive attributes. When Schumacher asked a sample of young adults what characteristics they associated with successful individuals, they consistently rated social and professional success with such attributes as tallness, strength, confidence, energy, cool-headedness and resilience.

Which brings me back to clothes, because our Victorian forebears always maintained that 'clothes make the man'. It seems that they were not so far off the mark here, because, in the US, Elizabeth Hill at Tulane University and Elaine Nocks and Lucinda Gardner at Furman University have been able to demonstrate that people's attractiveness is significantly affected by the clothes they are wearing. In tests, the same person wearing designer outfits and expensive jewellery was perceived as being of higher status and more attractive than when he or she was wearing more conventional clothes.

But why should appearance play so important a role in places where rational decisions are supposed to be the overriding concern? Well, it might have to do with the fact that we are constantly searching for cues that identify successful people. After all, anyone who can afford to buy smart new clothes can't have done too badly. Remember Salvador Dalí? Even as a penniless young painter, he insisted on living a lifestyle of flamboyant and conspicuous opulence far beyond his means: everyone thought he was doing extremely well because he was obviously attracting many wealthy clients, so they all came to him to commission paintings too. Success, you see, breeds success.

But why height? Why should successful people actually be taller than unsuccessful ones? Are tall people really that much better or is it that tall people inevitably overawe us so we expect them to be better? And, come to think of it, doesn't that bias things heavily against women when they are competing against men? Well, I'm not entirely convinced that the rules don't change when the two sexes are in competition. But if they are, it may just mean that you have to dress better than everyone else to get the Nobel Prize.

Voting for the tall one

OK . . . so Obama won the 2008 US presidential election. All that hard work on the campaign trail, and the several billion dollars spent by the various hopefuls over the year that the campaign lasted. It all paid off. We got the best man for the job thanks to the fierce winnowing effect of the democratic election process. A Darwinian triumph of selection for the best man.

Well, you might think so, but I'm not so convinced. Of course, it was something deeply embedded in our psyche and behaviour, honed by Darwinian evolutionary process over hundreds of thousands of years. But not quite in the way you might have imagined. In my view, science could have saved them all a lot of time and unnecessary money, at least in the endgame. McCain was set to lose come what may – and that wasn't just the Palin Effect.

In fact, the evidence was there all along, had the various campaign teams simply taken the trouble to ask the scientists. Obama was bound to win on two very simple grounds: he was the taller of the two candidates (the taller candidate has won three times more US presidential elections than the shorter candidate since 1900), and he had the more symmetrical face.

What on earth has facial symmetry got to do with it? And what's facial symmetry anyway?

Well, symmetry is simply being symmetrical, each half of the face being a mirror image of the other. It turns out that producing a nice balanced, symmetrical body is not as easy as one might imagine. Given all the vicissitudes of life – from illness to injury to starvation – during the long haul of development from conception to final adulthood, our genes have a hard time trying to build our bodies the way they were meant to be. It turns out that one of the markers of top-quality genes is how well they can cope with all these insults and still produce a symmetrical body. Facial symmetry (along with the symmetry of everything from breasts to fingers, foot length to ear lobes) is thus a rough and ready index of the quality of your genes – where by 'quality' is meant no more than the genes' ability to do their job and produce a functional body. It turns out that symmetry correlates with how well one does lots of things in life, but quite the most extraordinary and disturbing is that it seems to be a very good predictor of which candidate will win an election.

Tony Little and Craig Roberts, both then at Liverpool University, discovered that our voting patterns aren't always so carefully thought out as we imagine in our much vaunted democracies. Principles and plans are pitched against each other in the hustings, but it seems that's really just a smokescreen for the candidates to show off their bodies.

Little and Roberts first asked a large sample of people to choose which of two faces they would prefer to run their country. The eight pairs of faces were based on the actual winners and losers of the previous two national elections from the UK (Blair/Hague, Blair/Major), the USA (Bush/Kerry, Bush/Gore), Australia (Howard/Latham and

Howard/Beazley) and New Zealand (Clark/Shipley). Being just a wee bit canny, they did not show the actual faces, but instead showed the same, neutral face manipulated using fancy shape-changing software to have more or fewer of each of the two candidates' key facial features. These manipulated faces don't look like the originals, but they have their core physical features, such as lip and nose shapes, eye lines, cheek shapes and dozens of other barely noticeable traits. They produced two such faces, one based on the winners and the other based on the losers.

The outcome? Well, subjects chose the winning face nearly sixty per cent of the time, and the losing face only about forty per cent. More striking still, when they plotted the relative preference for one face over the other in the eight elections against the actual votes cast for that candidate or their party, they found a very good fit. Indeed, if preference was plotted against the actual number of seats won by each candidate's party, the fit was even better. So when they came to predict the May 2005 UK election, the experimental results based on the face preferences suggested that Labour (Blair) should win fifty-three per cent of the votes and fifty-seven per cent of the seats. In fact, on the day, Labour gained fifty-two per cent of the votes cast for the two major parties (Labour and the Tories) and sixty-four per cent of the seats won. That's pretty impressive.

But surely the voters must be taking note of all the promises and policies that the candidates and their parties make? Well, it seems not, for these results gel rather well with the remarkable fact that of all the US presidential elections since George Washington ascended the American throne where we have height data for the two candidates, the winner has been the taller in seventy-one per cent. Stature is another trait that appeals to us, and has many unexpected everyday consequences. There have been several studies in recent years showing that, statistically speaking, men's (but, it seems, not women's!) salaries are correlated with how tall they are. In fact, in the UK, your salary increases by about one per cent for every centimetre that you are taller than the average height for the population.

But I digress . . . because in a second experiment, Little and Roberts added a novel twist to their original experiment. They took the 2004 Bush–Kerry contest and asked a different set of subjects to say not just which face they preferred to run their country, but which they would prefer during time of war and which during time of peace. As before, they used a neutral face manipulated to have more or fewer Bush or Kerry's features.

The startling results were that the Bush-like face won hands down in the time-of-war condition (preferred by seventy-four per cent), but Kerry was the clear favourite in the time-of-peace condition (gaining sixty-one per cent of votes). Subjects were also asked to assess the two faces for various traits. 'Bush' was seen as being more masculine and dominant, whereas the 'Kerry' face was seen as being more attractive, forgiving, likeable and intelligent.

Which, you might say, is good news for Kerry. The bad news, it seems, was that he chose to run at just the wrong time, while the Iraq War was still in the forefront of the public's consciousness. Had he held off and waited until the following election (which Obama won), he might have done better. Was there perhaps a word of warning here for Hillary Clinton? Her naturally more feminine face might have stood her in good stead had the election been in the middle of a long period of peace. But, alas, the troops were still in Iraq and Afghanistan, and the rest, as they say, is consigned to the dustbin of history. Better luck on her choice of timing next time?

Of course, you might want to cite Abraham Lincoln as an obvious counter-example to the symmetry story. Poor boy, he got kicked in the face by a horse as a child, and grew up to have the most asymmetric face of any US president ever. Recent laser analyses of two plaster casts of his death mask reveal that the left side of his face was much smaller and thinner-boned than the right, hence his iconically craggy looks. Many people noted at the time that his left eye was inclined to drift, a further sign of a left-sided weakness. And it didn't seem to do him any harm in the political races of his day, did it?

Well, yes and no. There is one big difference between elections in Lincoln's day and those today: an image-based media. Photography was only just coming into its own at that time, and the best that most people got to see of their candidates was an artist's impression in a newspaper. It was not until well after the American Civil War (1861–5) that photographs became common in newspapers. Besides, Lincoln famously neither campaigned nor gave interviews during his presidential campaigns, but allowed his Republican Party election team to do it all for him. Very wise, you might think.

But the real issue is how he compared against his main rival, the Democrat Stephen Douglas. We don't know how symmetrical Douglas was compared to Lincoln – he could hardly have been less symmetrical, one imagines. But the one thing we can say is that Lincoln was much the taller of the two. Douglas, who was known to everyone as the 'Little Giant', was only five foot four, and a good twelve inches shorter than Lincoln who, at six foot four, was unusually tall for the time. With such a height advantage, symmetry was probably irrelevant. So, on the present hypothesis, Lincoln won fair and square. Case proven?

Politics? It's just physiology, dummy

The link between Lincoln and Douglas reminds me, rather serendipitously, of something else in similar vein. A recent study by Douglas Johnson and his colleagues at the University of Nebraska (which just happens to be in the small American Midwest town of Lincoln) sampled a group of people who had relatively strong political views – both right and left – for their emotional responses to threatening pictures. These included pictures of a massive spider on a very frightened person's face, a dazed bloody face and a wound covered in maggots.

They first divided their subjects into two groups: those who scored high or low on protecting the interests of their community from external threats – high scorers said they strongly supported military spending, searches without need of warrants, the death penalty, obedience, patriotism, the second Iraq war, school assembly prayers and the literal truth of the Bible, and opposed premarital sex, immigration, pacifism, gun control, gay marriage, abortion and pornography. Then, while they were looking at the pictures, they measured their subjects' physiological responses using both the galvanic skin response (the sweatiness of the palms) and the amplitude of the eye blink to a loud noise (an instinctive startle response). Those who scored high on the social-disciplinarian scale had a much stronger physiological reaction to these threatening pictures than to neutral pictures, compared to those with more liberal views.

In short, those who supported more extreme positions, especially on the political right, were more emotionally responsive kinds of people – in effect, more likely to panic when something untoward or unusual happened, more likely to react with a flight-or-fight response than a considered, rational one. Politics, it seems, is just an emotional response – as every demagogue from long before to long after Adolf Hitler has probably known only too well.

Perhaps not surprisingly, there was also an effect of education on these results. How long people had stayed at school correlated negatively with socially protective political views: the less schooling a subject had had, the more supportive he or she was likely to be of right-wing politics. But this effect was independent of the physiological response, serving merely to reinforce the physiological effect, not to explain it.

These particular physiological responses are probably associated with the activity of the amygdala, a relatively small, quite ancient part of the brain that processes responses to emotional cues in all mammals. Of course, it's perhaps not so much how your amygdala is tuned that makes you politically extreme, but that your intrinsic nervousness makes you more responsive to things that might seem to threaten your particular social world. Education probably plays an important role in dampening that response, by allowing the frontal lobes (where much of the brain's conscious work goes on) to counteract the emotional responses with a more considered view, so explaining why education is invariably the friend of liberal politics.

Twelve good men and true

One of the linchpins of British democracy has, of course, always been the jury system. Since medieval times, 'twelve good men and true' sit down and sift the evidence to decide the guilt or innocence of those hauled before the courts. So it may be no surprise that when the British government recently proposed to abolish juries in certain types of trials, the House of Lords – ever the guardian of ancient tradition, moral rectitude and privilege – roundly defeated the proposed bill. This set me wondering about the psychology of juries. For seven hundred years, the jury system – the right to be tried by your peers – has been sacrosanct under English law and all its derivatives around the world. But, given the number of cases whose verdicts have been overturned in recent years, I wonder whether you might prefer not to be tried by a jury next time the state has you in its sights.

The jury system was introduced, initially, purely for the benefit of their lordships. The deal (enshrined in Magna Carta, enacted at Runnymede in 1215) that the English nobles struck with the euphemistically named Good King John entitled them to be tried by their fellow peers rather than face summary judgement at the hands of the king and his rather dodgy henchmen. It was only centuries later that this right was extended to all and sundry (that's to say, the rest of us peasants).

So far so good. But think of the context in which such trials occurred. The population was tiny, and the twelve good men and true of the jury were drawn mostly from those with whom one lived. For most practical purposes, you really were being judged by your peers. In effect, when they were asked to decide whether you really did steal Old Mother Hubbard's shoe, they relied on their personal knowledge of you: were you *really* the sort of person who would do that? They probably didn't even need a trial to come to the right conclusion. OK . . . so sometimes they made value judgements and took personal sides, but you really were being judged by your community and what they found acceptable behaviour.

Today, however, it is all rather different. First, it's very unlikely the jury knows anything at all about you. Indeed, the lawyers insist on this, and will ask the court to reject jurors who have any personal knowledge of the defendants or the case. As the defendant, you might consider that an advantage, of course: better to have people who have no preconceptions deciding your guilt. But I wonder whether the community's interests are being so well served by this when mistrials mean hefty payouts from your taxes to people who have been wrongly convicted. And of course, hefty payouts to lawyers who get paid whether they win or lose, or even do a half decent job with the evidence . . .

A second problem is that forensic science is so much more technical now. Indeed, lawyers are often forced to simplify the evidence so that the jury can grasp its significance, creating yet more opportunities for confusion. This has turned out to be a particular problem in fraud cases, which often involve immensely complex financial transactions that need someone with the IQ of Einstein to understand.

Third, even cases of quite modest length become very taxing for the jury. In the world of the three-second attention span of television, the concentration required to keep track of convoluted legal arguments, complex evidence and the many strands of inference and innuendo that might be deployed by a good lawyer will inevitably tax most normal individuals far beyond their natural abilities. They simply cannot remember all the details. The reason is rather simple: decades of research in psychology has shown that memory for our experiences is not like a video tape. Instead, we remember a few salient features of what happened. When we are asked to recall what happened, we fill in the details and gaps on the basis of plausibility – what seems most likely to have been the case, given our everyday experiences – which is why witnesses commonly disagree about what they saw.

One last problem is the way lawyers work. The blunt truth is that lawyers do not exist to get at the truth, but rather to get the best deal possible for their clients, right or wrong. That means that they will always want to be as economical with the truth as they can. They are storytellers out to convince the jury to see the world from their point of view. In our legal system, the jury is passive and simply has to listen: they can't test the evidence for themselves, or question the lawyers' interpretations of the facts (perish the very thought, m'lud . . .). In my view, this is why mistrials are, relatively speaking, so common.

The last problem is the jury itself. Even once it is in the jury room, it is not twelve independently minded people trying to evaluate all the facts. Most juries are in fact juries of one or two people. One or two very forceful or highly educated individuals can often sway a jury by force of personality or their own competence in arguing cases. It's our evolutionarily well-honed psychology once again: the best thing for communities out on the plains is if everyone does the same thing, so a few good leaders and a lot of sheep is the perfect solution. There is no scope for bolshie individualists who ask too many questions. It's a real problem.

What's the answer? My suggestion would be professional juries: men and women who are qualified to understand the complexities of modern forensic science and complex arguments, paid to sit on juries as a job. The lawyers won't like it, that's for sure: they won't be able to bamboozle them so easily. But we may get fewer mistrials.

Chapter 14

Natural Minds

The question of what distinguishes us from other animals has probably exercised us for as long as we have been around as a species. It is not an easy question to answer, especially given that modern molecular genetics has been narrowing the gap with scant concern for human self-esteem. The one domain in which we still seem to stand apart, however, has been our minds. Human culture stands as one of the greatest of all evolutionary achievements. Our capacity for culture rests in part on our all but unique ability to introspect, to reflect on our own feelings and beliefs, and in particular those of others.

What's on your mind?

This ability to reflect on others' mind states is a capacity that children develop at around the age of four or five years, when, in psychologist-speak, they acquire theory of mind. A child aged three to four is a skilled ethologist: it knows how to manipulate others. Asked who has eaten the chocolate in the fridge, it knows that if it says in a very convincing way that it was the little green goblin from down the lane who hopped over the window sill, there is every chance an adult will believe it. But it does not really understand why this ruse works, and it certainly doesn't appreciate that the chocolate smeared around its face gives the game away. But with theory of mind in its mental toolkit, it knows how to manipulate others' beliefs about the world. Now, it can lie effectively. Suddenly, it has become a psychologist – it can read the mind behind the behaviour.

This capacity for theory of mind has been the great Rubicon that stands between us and the rest of the animal kingdom. Animals are stuck in the mental world of the three-year-old. But the question of whether other species share this capacity with us has continued to intrigue those who study the behaviour of animals. Do apes, genetically our nearest and dearest, share this unique trait with us? How about dolphins, or elephants? The problem that has bedevilled this area has always been how to design an experiment that unequivocally tells us whether animals share this trait with us. It is not as easy as it might seem.

However, a novel approach to this problem has been developed by two psychologists at the University of St Andrews. Erica Cartmill and Dick Byrne decided to let apes tell it their way. Instead of asking the apes to do experiments that required unnatural behaviour by the animals, such as pointing to where a reward might be hidden, they wondered whether apes could show that they understood mind states well enough to signal it in their behaviour. They used frustration from a thwarted outcome to trigger a response in orang utans.

The experiment was elegantly simple. They offered orangs the opportunity to beg for food from an experimenter holding two dishes, one containing a desirable food such as bananas, the other an undesirable food such as leeks. When the orang begged for food, it was given all the preferred food on one occasion, all the non-preferred food on another and half of the preferred food on a third. Then the experimenters waited to see what the orangs would do. They reasoned that if the orang thought that the experimenter had misunderstood their request, they would try a range of new gestures in an attempt to make the experimenter understand, but if they got half the desirable food they would repeat the same gestures on the grounds that what had worked partially first time ought to work again to get them the rest. And this is exactly what they found.

This is about as close as we have got to showing that apes can understand someone else's mind. If we must draw a Rubicon, then it puts the great apes on our side of the boundary fence. They are still not in the same league as adult humans, so they won't be writing works of fiction. But nonetheless, like us, they could imagine that the world could be other than it is. And asking that question, after all, is the basis of science. Everyone else has their nose pressed so hard up against the grindstone of life that they could not even entertain the thought.

Natural minds

We humans are naturally predisposed to attribute minds to other animals. It is simply a consequence of the fact that mind-speak is so deeply embedded in our everyday thinking. The philosopher Daniel Dennett referred to this as the 'intentional stance' – the tendency to assume that other individuals have minds like our own, ones that allow us to reflect (intuitively, even if not explicitly) on the contents of our own mind states. But what kinds of minds do animals have, and how do they compare with ours?

Psychologists have spent the past century or so exploring the mind in some considerable detail. In the course of this, we have learned a great deal about memory and learning, how animals solve problems or find their way around mazes. And the burden of all this effort seems to be that most animals are pretty much of a muchness in terms of these basic cognitive processes.

We have, I think, to be a bit dissatisfied with this conclusion. It's a bit like being handed a detailed summary of all the bricks, mortar, slates, wood and windows that make up a house, but without the breath of a mention of what the building itself looks like or why it's there. Or being given a detailed account of all the bits and pieces under the bonnet of a car, but not a word about how they function to propel the car along the road or why one might even want to do that. To me, that smacks a bit of trainspotting – making endless lists of engine numbers without taking the trouble to ask what trains are actually there for.

In fact, there is reason to think that at least some of the monkeys and apes are a bit different from the run-of-the-mill mammal and bird. It's their ability to handle social complexity that seems to mark them out, and this seems to depend on a peculiar kind of cognition that has come to be known as 'social cognition'. Monkeys and apes seem to differ from other animals in the intrinsic complexity of their social relationships. The important thing here is not that they can do certain kinds of behaviours that others cannot, but rather *how* they do them.

Primates engage in forms of behaviour that are unique

and do not occur in other non-primate species, as for example Dick Byrne and Andy Whiten's study of 'tactical deception' showed.* The important issue seems to be that they are able to appreciate how what they do will be misinterpreted by the other individual, and thus result in that individual behaving in a way beneficial to the actor.

The idea that monkeys and apes read minds (like humans do) rather than just behaviour (like all other species seem to do) has, however, faded somewhat with time. There is simply no evidence that any primates other than humans have a generalised capacity in this respect. Indeed, the only evidence for any kind of nonhuman mind-reading is from great apes. Even so, the evidence is not straightforward. While there is a lot of experimental evidence to show that chimpanzees can understand another individual's perspective, evidence that they have full-blown theory of mind is more equivocal. One study found that chimpanzees failed the critical kind of mind-reading task (the 'false belief' task) that young children pass with ease, while a second study showed that, although chimpanzees did do better than autistic humans (who definitively lack mind-reading capacities), they only did about as well as normal four-year-old children (who are in the process of acquiring mind-reading capacities, and so only have this skill imperfectly). It was this ambiguity that led Cartmill and Byrne to try a different tack with their orang utans.

Despite this, there is something very intense and personal about the social relationships of monkeys and apes that marks them out as very different from the kinds of relationships exhibited by other species. So far as I can

* See Chapter 3.

see, the only real exception in this respect seems to be domestic dogs, who seem to have been bred explicitly to exhibit the same kind of intense social commitment that primates have. Whether dogs' capacity to behave in this way is merely a superficial behavioural analogy of monkeys' capabilities or whether they produce these behavioural effects using the same kind of underpinning psychological mechanisms remains to be seen.

Nonetheless, these mind-reading abilities seem to give us some purchase on just what the differences between humans and other animals actually are. Intentionality is the capacity to reflect on the contents of one's mind, as reflected in the use of verbs like *suppose*, *think*, *wonder* (whether . . .), *believe*, etc. The capacity to use these words defines first-order intentionality: such an animal is capable of knowing its own mind. Most mammals and birds probably fall into this category.

More interesting are those cases in which the individual is capable of reflecting on someone else's mind state: *I suppose that you believe* . . . That capacity defines a higher level of intentionality, conventionally referred to as second order. It is equivalent to the stage that children achieve at about the age of five when they first acquire theory of mind. More interesting still is whether this sequence can be extended reflexively to yet higher orders. We have shown experimentally that normal adult humans can aspire to fifth-order intentionality as a matter of course, but that this represents a real upper limit for most people. Fifth order is the equivalent of being able to say: *I suppose* [1] *that you believe* [2] *that I want* [3] *you to think* [4] *that I intend . . .* [5] (with the successive orders of intentionality marked out in square brackets).

The hierarchical nature of intentionality provides us with a natural metric for scaling species' social cognitive abilities. If humans have a limit at fifth-order intentionality, chimpanzees (and perhaps other great apes) at second order, and monkeys at first order, then it turns out that these capacities are a linear function of the relative size of the frontal lobe of the brain (and only of the frontal lobe). This is interesting for two reasons. One is that the brain (and particularly the neocortex, the thin outer sheet that is both a mammal speciality and the seat of most of the complex behaviours we associate with 'thinking') has evolved from back (the location of the visual processing areas) to front. The frontal lobe is particularly associated with those capacities that psychologists refer to as 'executive function' (in very crude terms, conscious thought). Second, large neocortices in general (and large frontal lobes in particular) are a primate speciality, suggesting that whatever psychological capacities are underpinned by these neural structures are likely to be especially well represented among (if not unique to) primates.

So what are these capacities that monkeys and apes have?

In my view, it is not so much the capacities that differ between monkeys, apes and humans, but the scale at which each species can exercise these individual capacities. These capacities are in fact those basic to the lives of all mammals and birds. Minimally, they include the ability to reason causally, to reason analogically, to run two or more models of the world simultaneously, and the length of time into the future that any such model can be run. When these individual capacities are brought together on a large enough scale, mind-reading pops out as an emergent property. It looks like something special, and in one sense it genuinely is, but it is not some kind of specialised primate or even human capacity. Rather, it is the capacity to do better what everyone else does. In short, the differences between the various species of mammals on the scale of rats to humans is simply one of what might be termed the computational advantages of scale.

So limited a mind

Despite this, the mind that has given us poetry as well as modern science sometimes seems incredibly limited. One example of this is the fact that we so often seem to make do with simple dichotomies. We are 'for or against', 'on the left or the right', 'beyond the pale' (as opposed to within it), 'friend or foe'. And it's not just English-speakers that go in for these simple views. Like many traditional peoples, the San bushmen refer to themselves as *Zhu/twasi* – meaning 'real people' as opposed to the rest of them.

Which set me thinking. We seem to have an awful lot of these dichotomies in science. There is the well-known debate on the nature of light, for example. Is it really waves as the Newtonians supposed, or is it made up of particles (in the form of photons) as the quantum theorists argued? Then there was the great debate among the nineteenth-century geologists, between the 'catastrophists' and the 'uniformitarians'. The catastrophists followed the influential French taxonomist Baron Cuvier in arguing from the geological evidence that dramatic changes in the environment, such as floods and volcanic eruptions, led to the complete extinction of certain forms of life and their subsequent replacement by entirely new ones.

Uniformitarians, such as the eminent British geologist Sir Charles Lyell who was one of Darwin's mentors, insisted that the geological record showed gradual change, with a correspondingly gradual evolution of life forms.

Comparable debates occurred in physiology. In the mid-nineteenth century, the physicists Thomas Young and Herman von Helmholtz between them developed the familiar 'trichromatic theory' of colour vision, a view that gained credence from the discovery that the retina of the eye contains just three types of cells that respond to colour, one for each of the 'primary' colours (red, green and blue) identified by the physicists. A few decades later, however, the German physiologist Ewald Hering developed the so-called 'opponent colour theory' on the basis of experiments which suggested that the visual system perceives colours in terms of complementary pairs – blue/yellow and red/green.

What is perhaps more interesting than the dichotomies is the fact that the often vitriolic debates that have accompanied them were eventually resolved when someone pointed out that both theories were in fact right. Light does behave like both waves and particles on different occasions, and the choice may be as much a matter of analytical convenience as of underlying reality. Similarly, evolution does proceed at different rates at different times. Volcanic eruptions or comet impacts do force the pace by causing mass extinctions, but at other times evolution proceeds at a more leisurely rate with a steady turnover of mutations. And the two theories of colour vision turn out to apply at different levels within the visual system: the retina analyses light according to the three-colour theory, but the visual cortex does so according to the four-colour version.

Nor are these examples unusual. The way in which mammals perceive sound long provided the basis for an acrimonious argument between the 'place' and 'frequency' theorists. One group argued that the pitch of a sound is determined by how far up the organ of Corti the vibrations transmitted by the cochlea travel;* their opponents argued that it was the frequency with which the organ itself vibrated that determined the pitch. In fact, both theories are right: for good physical reasons, low-pitched sounds are analysed on the basis of frequency, while high-pitched ones are analysed on the place theory.

We've even had the same kind of disputes in mathematics. In 1764, the Reverend Thomas Bayes, an English Presbyterian minister and Fellow of the Royal Society, published a posthumous paper in which he sketched out a theory of probability based on confidence. It was an elegantly simple theory based on a single mathematical theorem that could be applied under any circumstances. But later mathematicians baulked at his ideas, preferring something that was more firmly rooted in observable facts: they argued that probability is better defined as something about the frequencies with which events (such as tosses of a coin) happen. Thomas Bayes and his theorem fell into obscurity. But the last laugh was on Bayes: it turns out that the frequency theory of probability is really just a special case of his theorem about confidence.

Then there is that old chestnut of 'nature versus nur-

* The organ of Corti is a highly sensitive membrane in the inner ear. The twenty thousand or so fine hairs that attach it to the fluid-filled compartment of the cochlea register the sound waves transmitted from the ear and turn them into nerve signals to the hearing centres in the brain.

ture', which reappears with such monotonous regularity as almost to count as a law of nature itself. And every time it reappears, it is resolved in exactly the same way. In the 1940s, nature versus nurture was the focus for the debate over the inheritance of IQ; later, in the 1950s, it resurfaced in ethology in the debate about the nature of instincts; then in the 1970s, it resurfaced in the even more vitriolic but no less muddled controversy that grew up around sociobiology. And it resurfaced again in the 1990s with the appearance of evolutionary psychology and the rather predictable response to that from the social sciences and parts of mainstream psychology. And each time someone eventually remarked that we cannot separate genetic from environmental influences in the development of organisms in so simple a fashion, even though, like waves and particles in light, it is sometimes convenient to talk of one to the exclusion of the other.

Our problem is that our minds just lack the intellectual capacity to deal with continua, especially if these continua involve the interaction of several variables operating along different dimensions. We are happiest with simple dichotomies because they save us having to think. Although evolution has no doubt provided us with a satisfactory rule of thumb for getting by in everyday life, thinking in dichotomies becomes increasingly unsatisfactory for handling the complexities beneath the surface that are the real stuff of science. Knowledge, it seems, is perpetually threatened by our own intrinsic limitations.

Well, I'm still waiting for some enterprising chemist to resurrect Joseph Priestley's phlogiston theory of combustion by showing that it is complementary to the oxygen theory that we owe to his arch rival, the Frenchman

Antoine Lavoisier. Lavoisier – who ended up on the guillotine for being one of Louis XVI's tax collectors – argued that things burn by consuming oxygen from the air, whereas Priestley (and pretty much everyone else at the time) claimed that when things burned they gave up a substance called phlogiston. Lavoisier used his skills as an accountant to show that things got heavier, not lighter, when they burned up, and so must have taken in something, not given it up, thereby paving the way for the modern atomic theory of chemistry. It's never likely to happen, of course; still, one has to wonder whether as great a chemist as Priestley could have been all wrong . . .

What's in a probability?

And here's another example of our sometimes distressing inability to think things through properly. Some years ago in the days before email, my Monday post brought a large brown envelope. To my surprise, it contained a request to take part in a chain letter. 'Send no money!' it said. Just send copies of this message on to five other friends and colleagues within four days and ask them to do the same. 'If you don't,' it went on ominously, 'bad luck will befall you.' Simple as that.

Well, being an unreconstructed empiricist of the old school, I was of course inclined to bin it. That I did not do so was simply because along with the letter came the accumulated correspondence that had passed successively down the line from its starting point in the USA. I started to read it out of curiosity.

What made these letters so interesting (every one from a professional scientist, by the way) was that they all desperately tried to prevent the sender being stigmatised as superstitious. 'Jim, you know I do not believe in this kind of crap,' pleaded one, 'but I am sending it to you anyway because . . .' Or, 'Ever since I was a kid, I just hated these chain letters and refused to send them on. But I'm sending you this one because . . .'

And what made the big difference? Very simply the threat of bad luck. Every single one ended with a plea for understanding: 'I've got a grant application pending, and I cant afford to take the risk . . .' or 'I've got a job interview next week, and with the job market the way it is right now . . .'

So, smiling to myself with an ever so slightly supercilious air of condescension, I put the bundle back in the envelope, and dropped it in the wastepaper bin. I had a busy week away from home ahead of me, with a conference to organise for the next day and the usual crises of the term looming.

Perhaps I ought to have recognised the signs sooner, but I didn't. The next day, Tuesday, my conference started on a bad footing because there was no extension cable for the projector and it was some way through the first (by then well delayed) session before one turned up. Wednesday and Thursday I managed to double-book myself on teaching arrangements for two different courses. Thursday, I reluctantly cut a meeting to make a lunchtime book-launch party on the other side of London, only to discover when I got there that I had turned up a week early. Returning home Thursday night, I discovered that my wife had taken to her bed with the flu. Then, as the weekend progressed, the rest of the family went down with it one by one, until finally it was my turn. It was a bad dose: I hadn't been ill enough to be off work for twenty-five years. The two boys were running temperatures of 103°F and it was the first time my daughter had ever had a day off in the eleven years since she had started school.

Now you know – and I know – that this was really just a long series of coincidences. But when you weigh up the probabilities of all five (or was it nine?) things happening in the same week, it does make you think, doesn't it? The odds must be around a million to one. Small wonder people start to believe in superstitions and astrology when things happen on this scale.

But if you analyse things more carefully, the odds turn out to be much less impressive. The domino effect of the flu on the family would have been more impressive if they had all been in different households, and that year's winter flu hadn't generally been acknowledged to be unusually virulent. Some of the classes at the children's various schools were down to half their usual size that week, and not a few families succumbed in their entirety.

Double-booking teaching is hardly unusual, especially in the first chaotic week of term. Nor is the late starting of a conference due to technical hiccoughs all that unusual. But wasting a lot of time chasing halfway round London to a launch party a week early – well, surely that's something altogether out of the ordinary? Yes . . . but I had actually written it in my diary as being that week when I received the invitation six weeks earlier – a long time before anyone had even thought of sending the chain-letter package to me, and possibly earlier than the whole silly chain had been started off. To include this in the calculation really would be cheating – or at least presumptive on the part of the Fates.

And if it comes to that, all of these events occurred before the four days of grace were up. In fact, it really was outrageously mean of the Fates to victimise me when I still had a day in hand to send on the letter and its contents! Nothing should have happened before Friday! Not one of these cases of 'bad luck' should count! In fact, history and hindsight tell me that, aside from the onset of my own dose of flu, nothing at all happened in the week starting with the fifth day after I received the letter.

So, the likelihood that all these mishaps were caused by my refusal to send on the chain letter was actually zero. In fact, the chances of something going wrong on any given day is probably quite high, though we tend not to notice most of them until something draws them forcefully to our attention. Then when something like a chain letter does raise them into our consciousness, we tend to look about for post hoc confirmatory evidence. Like I said – very unscientific.

Still, I suppose I shouldn't be too ungrateful, because the chain letter did set me thinking, and gave me a topic for an article that earned me the usual nice little fee . . . So, thanks, guys.

Chapter 15

How to Join the Culture Club

'I think, therefore I am,' declared the seventeenth-century philosopher-mathematician René Descartes, adding by way of afterthought that, since animals obviously didn't speak, they couldn't think and therefore certainly didn't have souls. We have lived in the them-and-us dichotomy of Descartes' shadow ever since. Nowhere has his influence been more intrusive than in the social sciences, where conventional wisdom has always insisted that the great divide between humans and other animals makes the latter totally inappropriate as models for the study of human behaviour. The great markers that set us apart from the brute beasts are culture and language.

The ever-moving goalposts

The argument, of course, hinges on the uniqueness of these two key phenomena. The result has sometimes been a near-farcical effort to defend the honour of our species against upstart claims that mere beasts might aspire to such a noble condition. Every attempt to show that some animal or other possesses language or culture has been met with a counter-claim that has tried to shift the goal-posts by redefining the terms. Man-the-tool-user rapidly became Man-the-tool-maker when it became apparent that many species of animals do in fact use tools.

So what is this culture of which we are so defensive? Half a century ago, the American anthropologists Alfred Kroeber and Clyde Kluckhohn surveyed the literature and emerged with some forty different definitions in current usage by anthropologists and social scientists. By and large, these seem to break down into three major classes of definition: culture consists of ideas in people's minds (social rules, patterns of ritual, beliefs, etc); culture consists of artefacts that are the products of those minds (so-called material culture like tools, pottery and its decorations, clothing, etc); culture is language and its products (high culture in the everyday sense, everything from Shakespeare to Bob Marley). The last, of course, brings us back to that other unique pillar of the human condition, language, and so rather shifts the goalposts again.

Apart from some inherent circularities (only humans have language, therefore only humans can have culture because culture is language), most of these definitions raise questions about the uniqueness of human behaviour. Are animals' minds really empty? Do they have *no* beliefs about the world? Are the hammers and anvils that the chimpanzees use for cracking nuts bona fide instances of material culture or not?

Bill McGrew (now at Cambridge University) has been a vigorous critic of the culture-as-artefacts school of human uniqueness. In his book *Chimpanzee Material Culture*, he challenged the advocates of this view to show why the chimpanzee's toolkit fails to meet the definitions they readily accept for humans. Three decades of intensive fieldwork in Africa has uncovered a long list of natural and manufactured tools that chimpanzees use, ranging from hammers to probes, fishing tools to sponges. Were we to lose the labels from such exhibits in a museum, he insists, we would be hard pressed to tell whether they had been manufactured by humans or apes. In only two respects does the chimpanzee toolkit differ from that of pre-technological human societies: chimpanzees do not have vessels of storage and do not construct traps (for fishing or hunting).

Two other widely touted examples of animal culture have long since entered into popular mythology. One is the way blue tits learned to remove the cardboard discs that once capped British milk bottles: during the 1940s, the habit of prising off the caps so they could sip the cream that (in those days) lay on top of the milk gradually spread among these little garden birds throughout much of southern England. The other is the habit of washing sand off sweet potatoes that spread through a troop of Japanese macaques once the habit had been invented by a young female named Imo.

Both examples have, however, received hard knocks at the hands of psychologists during the last few years. Several careful reconsiderations of the data have pointed out that, for a culturally learned behaviour, the rate of transmission through the population was remarkably slow in both cases. It took literally decades for Imo's potato-washing to spread to the rest of the troop; even then, only animals that were younger than her learned to copy the habit. The old dogs never learned new tricks. It seems that in most cases these new habits spread by a much simpler process: an observer animal's attention is drawn to a problem by the behaviour of the tutor, and it then learns the solution to the problem for itself by a process of trial and error. In humans, the tutor would teach the observer both the nature of the problem and the solution or the pupil would simply copy the tutor, and this marks a clear distinction between culture in humans and culture in animals.

Observations of this kind have led psychologists like Mike Tomasello, of the Max Planck Institute for Evolutionary Anthropology at Leipzig in Germany, to doubt whether any animal has true culture in the human sense. But before we leap to premature conclusions, we might bear in mind the questions that are being asked. Tomasello is interested in the mechanisms of transmission; primatologists like McGrew are interested in what the animals actually do. By any reasonable operational definition of culture, chimpanzees have culture, but, as Tomasello points out, we may legitimately doubt whether they learn it in quite the same way as we do. One way of asking the question, then, is to separate out the *capacity for culture* (apes can develop variations in behaviour that are random, casual innovations of no particular ecological relevance – a bit like wearing baseball caps backwards) but only humans have the *potential for culture* that allows them to exploit novel innovations which build progressively on what people have done before – the thing that made possible Isaac Newton's 'standing on the shoulders of giants' view of how science, a cultural activity if ever there was one, evolves.

Speak easy

It is obvious that what we often view as culture in humans is deeply embedded in language. We use language to describe, to teach, to intone our rituals. Animals, as Descartes observed, do not. Yet they are not dumb. Dogs bark, monkeys chatter. Conventional wisdom has always insisted that these are merely the direct products of the underlying emotions. Dogs bark because that is the kind of noise their vocal tract produces when they reach a certain level of excitement. While humans too produce similar kinds of vocalisations (screams and grunts), they also produce sound chains that are arbitrary yet meaningful. We can easily dismiss the much-vaunted waggle dance that honey bees use to notify each other of the direction and distance of nectar sources because it is specific to a very particular situation. Honey bees do not use the waggle dance to enquire after each other's health or sympathise over a misfortune.

Yet, recent research suggests that, when it comes to monkeys and apes, it may be necessary to turn conventional wisdom on its head. Dorothy Cheney and Robert Seyfarth, of the University of Pennsylvania, carried out a series of ingenious experiments on wild vervet monkeys in Kenya's Amboseli National Park. By playing vocalisations of known individuals from hidden speakers, they have been able to demonstrate quite uncontroversially that vervet vocalisations convey considerable information that is quite independent of the behaviour of the vocalisers. Vervets reliably use calls to refer to specific kinds of predators (leopards versus birds of prey versus snakes). They know from minor differences in sound whether a grunt is a comment on what another vervet is about to do or on something it has seen, such as whether the caller is being approached by a dominant animal or a subordinate. In their more recent work in Botswana, Cheney and Seyfarth have demonstrated that baboons use grunts in a way that amounts to an apology in order to mollify an ally they have previously offended. And all this with what was once thought to be a simple all-purpose grunt.

There is, it seems, much more to animals' vocalisations than we had supposed. Like the proverbial visitor to China, the naïve observer hears only a jumble of sounds where in fact something much more complex is going on. We have been, and still are, mere beginners when it comes to deciphering the languages of other species.

More impressive still are the achievements of the language-trained chimpanzees, which I'll discuss in more detail in Chapter 21. Around a dozen chimpanzees, a gorilla and an orang utan have now been trained to use a variety of artificial languages, and the chimps in particular have demonstrated quite remarkable abilities, responding to instructions and answering questions at the cognitive level of young children. More alarmingly, perhaps, most of these achievements have been matched by an African grey parrot, the late and much-lamented Alex, who used spoken English to communicate.

Cogito ergo . . .?

There remains, however, one crucial stumbling block for animals. The ability to engage in the higher forms of culture that we associate with religious ritual, literature and even science depends on the ability to step outside oneself to see the world from an independent perspective. This requires being able to ask not just 'What happened?' but also 'Why did it have to be that way?' Animals, it seems, take the world as it comes. Only humans seem able to detach themselves from their own parochial concerns to imagine that things could be other than they are. Only then is it possible to ask the all-important 'Why?' questions that adults find so infuriating in children.

In the social context, this ability to stand back from the way things are is referred to as possessing a 'theory of mind'. It underpins our ability to understand another person's beliefs and the way we use this knowledge to exploit and manipulate each other. Children do not possess it at birth: they acquire the ability at around four years of age. In fact, some humans (such as autistic people) never acquire it. Neither sophisticated lying nor fictive play are possible until a child has acquired theory of mind. Without it, fictional literature is impossible and both science and religion, with their need to imagine impossible worlds, are out of the question.

It is equally clear that no animals reach this exalted state of mind. Monkeys can, of course, engage in deception, but it is deception of the kind that three-year-old children are good at. They can read another's behaviour well enough to exploit them, but they cannot understand that another individual can hold beliefs that are different from their own. The only exception, yet again, seems to be the great apes, as we saw in the previous chapter.

The substantive point is surely that the continued insistence that culture is a phenomenon which sets humans apart from the rest of creation seems to smack more of generic chauvinism than anything else. There are, of course, aspects of human culture that are not found in other species, just as there are aspects of language that appear to be unique to humans. These are but the high points on what in reality is a continuum. And therein perhaps lies part of the problem: humans seem to find it extraordinarily difficult to think in terms of continua, preferring instead to deal in simple dichotomies of them-and-us. We should recognise that neither language nor culture are simple unitary phenomena and that we share many of the processes that underpin them with at least some of our fellow creatures.

Why Shakespeare really was a genius

One thing, however, does seem to be uniquely human, and that is the fictional world. Animals simply could not understand what a story was – not just because they lack the language to understand the words, but because they are unable to comprehend the whole notion of imaginative fiction. If they did have language, they would take the story at face value, and be utterly perplexed by statements about a world that did not exist.

This is obvious if you think about William Shakespeare sitting down to write his play *Othello*. He has three core characters: Othello himself, Iago and the ill-fated Desdemona. To make the play work, he must persuade his audience (when they eventually get to see the play) that Iago *intends* that Othello should *believe* that Desdemona *is in love* with someone else. That involves three separate mind states on the stage. But to make the story really convincing, he has to add in Cassio, the apparent object of Desdemona's desires. If Desdemona merely fantasised about Cassio,

Othello would, surely, have been much less bothered by it all. It might have led to a bit of leg-pulling in the garden, but why otherwise should Othello be so exercised about the intelligence that Iago offers him – unless he is led to believe that Cassio reciprocates Desdemona's interest? It is this that racks up the intensity of Othello's angst and causes him to do what he eventually does. So, to make the story really sell, Shakespeare has to show or imply four mind states: Iago *intends* that Othello should *believe* that Desdemona *loves* Cassio and Cassio *loves* her.

But this is not the end of the story, because Shakespeare has to persuade the audience to believe all this stuff. If they are not taken in by it, the play will be dead in the water. So Shakespeare has to factor the minds of the audience (or, at least, the virtual mind of a nominal member of the audience) into his calculations. And last, but not least, he has to be doing the imagining of all this himself. So when he sits down with his quill pen poised above a sheet of foolscap one wet Monday morning in Elizabethan London, he has to be able to work – minimally – at sixth-order intentionality: he *intends* that the audience *believe* that Iago *wants* Othello to *suppose* that Desdemona *loves* Cassio and he in turn *loves* her.

That's no mean feat, because he is already working at one level of intentionality above what the average adult human can cope with. Notice that he is also pushing his audience to its limits – they are having to work at fifth-order intentionality. It is probably precisely because Shakespeare could work successfully at this level, and so challenge his audience to their limit, that he came to be such a successful playwright.

However, the real issue for our present concerns is that

only a human could have done this. With its cognitive limits set firmly at second-order intentionality (at best!), even the proverbial chimpanzee sitting down at its typewriter could never have produced the script for *Othello*. If it had actually done so after many millions of years of typing, it would have been a purely statistical accident, and not a very interesting one at that. For the ape typist would not have *intended* the action of the play, and it certainly would never have pondered the audience's capacity to follow the unfolding story as it did so. It might have appreciated that Iago intended to say something to Othello ('I *believe* that Iago *intends* . . .'), but it would not have been able to understand how, in addition, Iago intended Othello to interpret his words – that would have required third-order intentionality that it could never aspire to.

So the lesson for us is that the flights of fancy that we engage in when dabbling in literature, even when just telling stories around the campfire, are far beyond the cognitive capacities of any other species of animal currently alive. Great apes might be able to imagine someone else's state of mind, and so they might even be able to construct a very simple story, but it could never be much more than a narrative involving one character. Only adult humans could ever intentionally produce literature of the kind that we associate with human culture. It is possible, of course, to produce stories with third- or even fourth-order intentionality (perhaps the equivalent of the cognitive abilities of eight- and eleven-year old children), but they inevitably lack the sophistication of the stories told by the average adult, never mind those of a Shakespeare or a Molière.

More importantly, to really be able to challenge and

fire up the audience, a great story-teller has to be able to take the audience to the limits of *their* intentional abilities at fifth-order intentionality. But that means that the story-teller has to be able to work at least one level higher at sixth-order intentionality. That is beyond the scope of more than three-quarters of the rest of us. Shakespeare really was a genius.

Chapter 16

Be Smart . . . Live Longer

At root, it is, of course, our intelligence that has made us what we are – one of the most successful species ever to have lived (well, if we don't count most beetles anyway, given that forty per cent of all the animal species that have ever been described are beetles . . .). But to be fair, without our remarkable capacity to think through problems while building on the accumulated knowledge of the past, we would not have colonised every continent on earth, built the Great Wall of China, discovered radium, composed Bach's cantatas and Mozart's operas, landed men on the moon or devised the internet. In fact, being smart has all sorts of unexpected consequences for us and we shouldn't knock it. IQ is good for you.

Be smart . . . live longer

If you were born in Scotland in 1921, the initials IQ might just prompt you to remember Wednesday 1 June 1932. It was not a day of particularly high drama: no cup tie saw crowds streaming to one of the great football stadiums, no unexpected summer storm lashed the Western Isles, the Forth Bridge did not collapse. In fact, it was quite an ordinary day as summer days go. But that day you took part in something that was quite unique. Instead of the usual joys of school, you were taken off to some draughty hall to sit an intelligence test. Perhaps the recollection of it is only hazy now, lost beneath the memories of life's more important ups and downs. But think back for a moment, and reflect on the fact that, on that day, you took part in a remarkable experiment. Scotland's entire cohort of schoolchildren born in the year 1921 sat that exam with you – a complete and unique record of a country's scholastic abilities at one particular moment in time.

And you will probably be glad to know that, after all these years, your earnest struggles with pen and paper that day have not gone unrecognised: they have become a goldmine for researchers. One of the most remarkable findings to emerge has been a link between IQ, health and death. Indeed, if you are reading this now, it is in part because you were among the smartest of the children born in 1921. Of course, we have known for a long time that intelligence, health and mortality are related to each other, but we have always supposed that the link was indirect – through social deprivation and educational opportunity. Now a major study led by Edinburgh University's Ian Dearie has discovered a more direct link between IQ at age eleven and your chances of celebrating your eighty-fifth birthday.

Showing this was not an easy task. Dearie and his team had to track down the vital records of the individuals who took part in the original study, matching up the records of death so that they could determine who had died and who was still alive. An earlier study based on a sub-sample of 2,800 Aberdonians provided the first evidence that

IQ affected your chances of surviving into your seventies. But it was impossible with these data to separate out the effects of social deprivation from those of IQ. Then someone remembered that there had been a follow-up study during the 1970s of a cohort of people living in Paisley and Renfrew who had sat the IQ test in a second study in 1932. The follow-up study had focused on health, employment and levels of deprivation. From the Paisley/Renfrew study, they were able to locate 549 men and 373 women who had sat both the Moray House IQ tests in 1932 and the 1970s mid-life health check, and whose lives in the subsequent quarter-century could be tracked through the national records.

IQ is standardised at a notional value of 100 as the average for the population as a whole, with around two-thirds of people having an IQ between 85 and 115. Ian Dearie's analyses of the data from the 1932 Moray House study revealed that when socio-economic class and deprivation were controlled for statistically, each point drop in IQ at the age of eleven corresponded to an extra one per cent chance of dying before the age of seventy-seven. For someone at the bottom edge of what we usually consider the 'normal' range (i.e. IQ = 85), that meant that their chances of celebrating their seventy-seventh birthday were fifteen per cent lower than someone with an IQ of 100.

The effect was much stronger in lower socio-economic groups than it was among the better-off families, reflecting the well-known effects that economic deprivation have on health. However, this makes it clear that social, educational and economic deprivation alone are not the causes of IQ-related mortality, though they each obviously have some effect. Rather, the causes must lie in something more organic.

The most likely explanations are either that IQ is an index of early developmental factors or that it provides us with a general measure of what we might think of as 'organic integrity' – the effectiveness with which all the body's systems work. We now know, for example, that your experiences in the womb influence your chances of coronary disease and the risks of dying from heart attack or stroke later in adult life. We also know that these risks are associated with your birth weight, which is itself partly a reflection of your experience in the womb. We also know that low birth weight affects childhood academic abilities, and IQ more generally.

The intelligent butterfly

The film *A Beautiful Mind* paid tribute to the genius, if also the troubled mind, of John Nash, discoverer of the Nash Equilibrium in mathematics and winner of the 1994 Nobel Prize for Economics. But what the headlines don't tell us is whether behind the beautiful mind there was also a beautiful body – and not just that of Russell Crowe who played Nash in the film. In fact, it has always seemed to me that not all the swots I knew at school and university were dull, ugly or uncoordinated. Many were body-beautiful and not a few excelled in sports.

It now seems that there may be more to this than mere hearsay. Tim Bates, a psychologist at Edinburgh University, has recently shown, in a sample of over 250 people, that there is a small but significant correlation between IQ and bodily symmetry (based on the left-side/right-side symmetry of finger, hand and ear length). Symmetry is one of the components we recognise as beauty. So it seems that beautiful people are, on average, more intelligent, even though – as with all things biological – lots of other factors intrude to affect any given individual's performance.

Alas, there are knock-on consequences of this. Not only is it a well-established fact that taller people are more successful in social and economic life – on Wall Street and in the UK financial markets, taller people earn more, even when doing the same job – now it seems that the same correlation holds with IQ: several recent studies have demonstrated a correlation between IQ and success in the adult world. One study used a longitudinal sample of American baby-boomers (in this case, the cohort born between 1957 and 1964, representing the tail-end of the spike in births that followed the end of the Second World War). It found that each point increase in IQ added between $234 and $616 to income (though that didn't necessarily affect gross wealth). Other studies yielded similar results, but have also found an additional effect due to parental socio-economic status. It evidently pays to pick your parents carefully, but if all else fails it seems you can still haul yourself up by your bootstraps if you are smart enough.

However, to add insult to injury, it seems that not only do the beautiful get to be richer, but they are actually more fertile. Some years ago, my Polish colleague Boguslaw Pawlowski from the University of Wroclaw and I used a large Polish medical database to show that tall men were not only more likely to be married, but also had more children. In terms of evolution, they had higher fitness – made a greater contribution to the species gene pool – than shorter ones. Daniel Nettle, of the University of Newcastle, later showed much the same effect in a longitudinal British sample that had been studied since birth (and who were, at the time of his study, in their fifties, and so had completed most of their reproduction).

We had thought that this was simply because tall men are more attractive, and so are more likely to find partners and have babies. However, it now seems that the beautiful are also more fertile. Ros Arden and her colleagues from King's College London have recently shown, using an American military sample, that symmetry correlates with sperm count and sperm motility. Beautiful people are just more fertile. Life just isn't fair.

Mens sana in corpore sano

It used to be said of a certain Oxford college during the 1960s that its dons assessed prospective undergraduates by throwing rugby balls at them as they came into the interview room. A fumbled ball meant the thumbs down, a drop kick into the wastepaper basket an instant scholarship. Such selection practices were, of course, frowned upon by the sniffier colleges.

Yet I seem to recall that, in terms of performance in the academic league tables, that particular college by no means disgraced itself compared to other colleges that pursued more orthodox methods of selection. In fact, the critics ought to have been completely silenced by the results of a long-term study of educational achievement published back in the 1970s. It revealed that the typical high-flyer was not the conventional bespectacled genius of Billy Bunter's Greyfriars, but the all-rounder. High-flyers, it seems, tend to fly high in everything from sports to exams

– and just to add insult to injury, even the social sphere is not excluded.

No doubt this slightly surprising result in part reflects the fact that there is nothing like success to breed success. But I wonder whether there isn't also something in the old educationalist's adage that healthy minds are found in healthy bodies – *mens sana in corpore sano*. This is not to say that sporting types can be intellectual geniuses simply by virtue of being sporty. But a heavy involvement in sport might provide one essential ingredient for being able to make the grade intellectually. The reason may simply have to do with one of today's endocrinological buzzwords – endogenous opiates.

The endogenous opiates, or endorphins, are the body's own painkillers. They are pumped round the brain in vast quantities whenever the body is subjected to stress, thus buffering us against the pain of tissue damage. This system is presumably designed to allow the body to continue functioning more or less normally when a failure to do so because of injury might result, say, in the animal being caught by the predator. But what have painkillers got to do with intellectual activity? The answer perhaps lies in the fact that we often refer to it as intellectual *effort*.

A curious myth has been perpetuated over the centuries to the effect that geniuses produce works of genius effortlessly. René Descartes was partly to blame for this. He affected the lifestyle of a dilettante and habitually spent most of his day in bed while churning out works of genius in the afternoons. T. E. Lawrence (of Arabia fame) did his bit too, claiming to have attended no more than a dozen lectures during his entire undergraduate career before gaining an effortless first-class degree at one of the better Oxford colleges (Jesus).

But my impression is that these kinds of claims are ninety-seven per cent bravado. They invariably conceal a great deal of very hard work behind the scenes – often in the college library. Lawrence's renowned knowledge of medieval crusader castles (he wrote a seminal report on one excavation in Palestine) was not acquired by divine inspiration. And my guess is that Descartes was doing a great deal more than dozing as he lazed in bed each morning. What he was in fact probably doing was exactly what every good mathematician still does – namely, allowing his subconscious to mull over a problem off-line.

Which brings me back to opioids. What they surely provide is a buffer against the pain and stress caused by the physical and mental exhaustion, the discomfort and eye strain, headaches and frustrations that come from poring over books, other people's obscure algebraic proofs, and experiments that refuse to turn out right. Those lucky individuals with naturally high endorphin levels sail through all this and emerge at the other end fresh and still raring to go long after other mere mortals have wilted and given up.

Now, one way of raising endogenous opiate levels is to exercise vigorously on a regular basis. Of course, I don't want to suggest that exercise will turn everyone into a genius. Clearly, a certain amount of native intellectual competence is required – things such as memory and quick logical thought, which usually come under the rubric of general IQ. All I am suggesting is that we may have overlooked an important element in the equation for that multifaceted trait we refer to as IQ, namely endurance. Those who have the cerebral machinery will not succeed unless they also have the capacity to stand up to the work effort required to exercise it to the full.

Which raises some interesting questions. Should lectures begin with ten minutes of advanced callisthenics before getting down to the business of working through the proofs of matrix algebra? Do field workers in biology who spend their days tramping the moors have an unfair advantage over their more sedentary colleagues in, say, English literature? Should a high endorphin titre in the brain count as an essential qualification for an intellectually stressful job? Should prospective employers have a keener interest in the kinds of exercise you take – or don't take?

Perhaps next time someone else gets the job you desperately wanted, you should ignore their paper qualifications: instead, try checking out the way their muscles ripple under the well-cut outfit as they walk into the interview room.

And we might well contemplate the implications of this for how we educate our children. Physical sports have gradually dropped off the list of activities that children are asked to indulge in, partly through some rather odd notions about equality (the 'everyone should get a prize' mentality), but also, in these increasingly litigious times, partly because of the abject terror of being sued that turns both schools and local councils into quivering wrecks. But if there really is a relationship between exercise and learning, that might not be too clever, because everyone ends up suffering thanks to the stupidity and greed of a few. The real issue is that we need to learn how to accept risk and be less petulant and blaming when accidents happen. Life is full of risks, and you can't be grateful for the enormous benefits that invariably accrue from taking them and then blame others when it goes wrong – a lesson apparently lost on the world's bankers. Failure to appreciate that is a form of shortsightedness that doesn't do our children any good in the long run.

It still pays to learn

Despite all its inbuilt advantages, just being smart is not enough. Having the IQ of Einstein is a bit like having the biggest computer ever built: that's all very impressive, but without the software it's going nowhere. Education remains the key ingredient. Without packing the mind with knowledge and skills for it to mine and exploit, native IQ alone wont get you all that far. Education allows us, in Newton's famous phrase, to stand on the shoulders of the giants of the past. Knowledge, and especially scientific knowledge, is cumulative.

So, given all the later conflict between science and religion, it is all rather ironic that one of the most successful experiments in education ever done was actually carried out at the behest of religion – in this case, the Calvinist Presbyterians in Scotland. The impetus to ensure that every crofter could read the Good Book for him- (or even her-) self produced, by the early nineteenth century, what was probably the best educational system in the world. Literacy rates in Scotland were on the order of seventy per cent by the end of the eighteenth century, at a time when they were not much more than half that in England and Wales, never mind the rest of Europe.

By the mid-nineteenth century, attendance at university was more than ten times higher per head of population in Scotland than it was in England and Wales. And where higher education remained the near-exclusive preserve of the upper classes in England, it was the broad egalitarianism of the Scottish educational system that was its great achievement. Crofters' sons had virtually the same chances of making it to university as the sons of the laird and the minister. Education became a passport to a better life for droves of Scots, even though many of them went abroad to administer, explore, industrialise and generally create a virtual empire around the world.

The downside, of course – and this is not always recognised – is that all this education was probably responsible for almost as great a depopulation of the Highlands and islands as the Clearances themselves. OK, in this case, at least, this was seen as a good thing by the families – a way out of grinding poverty, a gateway to a future that was always going to be better and more sustainable than the harshness of life on the land back home.

That enthusiasm for buying into the educational dream had one important consequence. And that was an intellectual interest and curiosity right at the roots of society. One need only point to Robert Burns's father who anxiously sought out an education for his children (and how much less rich the world of literature would have been had he not!). It spawned what became known as the Edinburgh Enlightenment of the late eighteenth century when the philosopher David Hume and the economist Adam Smith and their friends rose from humble beginnings to write some of the most lastingly influential works of all time. It generated some seminal contributions to science, engineering and literature in the nineteenth and early twentieth centuries – names like Alexander Fleming, Walter Scott and the various Stephensons of railway and iron bridge fame.

Somehow we have lost that sense of purpose. Education no longer seems to be valued for itself, something to challenge the mind, to excite and motivate a spirit of enquiry. I do not know what the answer is, but I do know that unless we can find an answer quite quickly we are heading for deep trouble. The problem is summed up for me by the fact that applications for science courses at British universities have been declining at a steady rate for the better part of a decade. When I analysed the figures for chemistry and biology a few years ago, the decline was so precipitate that, if it continued at the same rate, the number of applicants for *both* disciplines would hit zero by 2030.

But my real concern is this. An education is not just a technical training in the arcane knowledge of a discipline (whether that be history, politics or a science). It is a training in how to think and evaluate, how to marshal evidence for and against a position, how to approach a problem critically without falling prey to prejudice and preconception. Those are skills that everyone from bank manager to politician, journalist to local government functionary, needs every working day. But to train those skills, it is necessary to excite an interest. And somewhere along the line between primary school and university, we are managing to stamp out that sense of excitement and enquiry. We will rue the day we lost sight of that.

Chapter 17

Beautiful Science

Polymaths of science

In a Gallup poll commissioned by the BBC some years ago, eighty per cent of the British people thought that science was important. That's pretty encouraging, isn't it? Well, yes, except for the fact that, by implication, twenty per cent of the population took a distinctly more jaundiced view of our activities. This is a figure that accords well with many other polls: typically, five to twenty-five per cent of the people polled express negative attitudes towards science.

So who are all these Doubting Thomases? And do they really matter? As a matter of fact, I think they do matter – very much. For their position within society often gives them an influence over our future history that far exceeds their numerical share of the vote.

By and large, people who are disdainful of science are well-educated, professional people. Typically, they hold degrees in the humanities: some are teachers, some are academics, others are members of the artistic and literary communities. More worryingly, some are politicians. They share a common antipathy towards science that is generally founded on the view that scientists are acultural and insensitive to the finer things in life. The underfunding of the arts relative to the sciences is regarded as symptomatic of this – our cultural heritage being eroded and submerged beneath the harsh adamantine machinery of science.

This is very much the Victorian caricature of scientists: the mad Dr Frankenstein hell-bent upon world domination even at the expense of his own life; the evil duplicity of Dr Jekyll. Whatever happened, I wonder, to Renaissance Man, that intellectual polymath whose interests ranged from music and poetry to astronomy and physics, and whose accomplishments and reputation often rested as much on the ability to turn a fine sonnet as on the construction of some ingenious experiment?

One thing seems clear: Renaissance Man is no longer always to be found among the humanities. A surprising number of scientists turn out to have hidden (and in some cases not so hidden) talents. Take Einstein, surely the archetypal scientist. Like many mathematicians, he was an accomplished musician: he played the violin. He was not, of course, a Yehudi Menuhin, but he did on more than one occasion play with a celebrity orchestra. Still, if you want to be sniffy about Einstein, then try Alexander Borodin, the nineteenth-century Russian commonly credited with having been one of the technically most innovative composers of his day. He taught chemistry for a living throughout his working life.

Speaking of chemists, I'm reminded of that other great Russian genius, Alexander Solzhenitsyn. After taking a degree in mathematics at Rostov University, he taught physics and chemistry before turning his hand to writing the novels that made him famous. And why should the eastern Europeans have all the credit when Britain has its own C. P. Snow, who, despite the disadvantage of having been a research physicist at Cambridge and, later, scientific adviser to the British government, went on to establish an enviable reputation as a novelist during the 1940s and 1950s.

Nor need we look so far back in time to find eminent scientists at work in the literary and artistic domains. Many will know that the astronomer Patrick Moore was an accomplished performer on the xylophone, an instrument for which he also composed.

On the literary side, we have zoologist John Treherne who, after publishing two successful books of historical biography (one on the iconic American gangsters Bonnie and Clyde), went on to produce a couple of well-received novels. His last novel, *Dangerous Precincts*, was an historical study of ecclesiastical intrigue and scandal set in the 1920s. And what about Richard Feynman (of *Surely You're Joking, Mr Feynman?* fame): wit, raconteur, sometime poet – oh, and yes, Nobel laureate in physics too. Not to mention, of course, a long line of widely acclaimed writers of science fiction from Isaac Asimov to Arthur C. Clark. And then there is the reproductive biologist and TV personality Robert Winston: early in his career, he dropped out of science for a few years and became a theatre director, winning in the process the National Directors' Award at the Edinburgh Festival in 1969.

Come to think of it, even among my own inevitably limited circle of professional acquaintances I can think of at least half a dozen scientists who perform regularly in music groups – two in chamber orchestras, one in a consort of viols, another in a madrigals ensemble, while the fourth, a clarinettist, is in constant demand for local jazz bands. Three others earn a pound or two as artists or illustrators (one now professionally). And they all manage to do this while working as academic scientists.

But perhaps it is only fitting that the final honour should belong to physicists. In 1987, the prestigious Cleveland Orchestra under its then principal conductor Christoph von Dohnányi gave the world premiere of the latest work by the American minimalist composer Philip Glass. It was a piece entitled *The Light* and had been commissioned to celebrate the achievement of two local boys, Albert Michelson and Edward Morley, exactly one hundred years before. Now known to every physics student as the Michelson–Morley experiments, their work had finally put paid to the then received wisdom that space is filled with an ether through which celestial bodies and such phenomena as light travel, so paving the way for Einstein's theory of relativity just two decades later. When science itself commissions art, it surely ceases to be philistine.

So it seems to me that Renaissance Man is very much alive and well. But if you want to find him or her, you probably shouldn't go looking in the nearest humanities department. Just try looking across the laboratory bench right across from you.

Poets can be scientists too

We don't often associate poets with science, but it seems to me that what distinguishes a great poet from a mere rhymer is much the same as what distinguishes a good scientist from the merely mediocre – an acute power of observation and that capacity for introspection that underpins human culture in all its forms. Take Robert Burns, that greatest of all Scots poets – whose two hundred and fiftieth anniversary we also celebrated in 2009, as it happens. To be sure, Burns was incredibly well read, especially for a 'humble ploughman'. Nonetheless, it is unlikely that he gained much by way of an education in even the rudimentary sciences of the mid-eighteenth century under the tutelage of his early teacher, John Murdoch. Nor, when Burnes senior (he changed the spelling of the family name when his children were born) took over his sons' education after Murdoch moved on to financially more rewarding things, would he have gained all that much from such books as William Derham's *Physico-Theology* and *Astro-Theology* which Burnes Senior borrowed from the local branch of the Ayr Book Society.

Indeed, Burns was notoriously unimpressed by the educated kirkmen of his day, with their book learning and lack of commonsense. As he remarked,

What's a' your jargon o' your Schools,
Your Latin names for horns an' stools?
If honest Nature made you fools,
What sairs [says] your grammars?
Ye'd better taen [taken] up spades and shools,
Or knappin-hammers.

In other words, get a proper job and do some farming or navvying. Or, on the virtues of two giants of the Scottish Enlightenment, the economist Adam Smith and the philosopher Thomas Reid:

Philosophers have fought and wrangled,
An' meikle [much] Greek an' Latin mangled,

Till, wi' their logic-jargon tir'd,
And in the depth of science mir'd,
To common sense they now appeal –
What wives and wabsters [weavers] see and feel!

All this intellectual effort, and you just tell what every fishwife already knows from folklore.

Burns may not have speculated deeply on the planetary spheres, the nature of light or the transmutation of metals, but he did give us some scintillatingly acute observations on psychology. Forget his 'To a Louse' – you need look no further than his wonderful narrative poem 'Tam O'Shanter' to find what, to my mind, are two of the most perceptive lines ever penned. As the poem opens, Tam sits 'bousing' in the alehouse with his friends, squandering his meagre takings from market day on booze. Meanwhile, back at home:

. . . sits our sulky sullen dame [Tam's wife],
Gathering her brows like the gathering storm,
Nursing her wrath to keep it warm.

One can point to observations that, while undoubtedly coloured by an element of self-interest on Burns's part, turn out to be solid science:

Let not women e'er complain
Fickle man is apt to rove!
Look abroad thro' Nature's range,
Nature's mighty law is change.

It is one of the cornerstones of contemporary evolutionary biology that, because of the way mammalian reproductive biology is organised, male mammals are naturally predisposed to polygyny. Only in those cases where males can invest directly in the business of rearing offspring do they opt for monogamy. Consequently, monogamy is rare in mammals outside the dog family: ninety-five per cent of mammalian species mate polygamously.

Worse luck for Burns, perhaps, humans happen to be one of the exceptions, mainly because, in our case, the business of rearing extends far beyond the moment of weaning, allowing males to invest in the processes of socialisation as well as through the inheritance of accumulated family wealth. Of course, human monogamy is not the kind of eternal, unswerving commitment that we often associate with swans and many birds. In contrast to mammals, ninety per cent of bird species have a monogamous breeding system, as Burns himself notes:

Among her nestlings sits the thrush:
Her faithfu' mate will share her toil . . .

Mind, to be fair to Burns, the wonders of modern molecular genetics have revealed that, even among supposedly monogamous birds, extra-pair matings are surprisingly common. Indeed, it is by no means impossible for every egg in a clutch to have been fertilised by a different male, even in pair-forming species. It turns out that a female bird can store sperm from different males, and selectively draw on it to fertilise her eggs when she is ready to lay them.

But there are a couple of remarks that Burns makes which are especially striking, not least because they make claims that have been explicitly demonstrated to be true only within the last decade. One is the fact that we can only sustain a limited number of friendships at any one time (see Chapter 3). Burns alludes to this in his 'Epistle to J. Lapraik':

Now, sir, if ye hae [have] friends enow [enough],
Tho' real friends I b'lieve are few;
Yet, if your catalogue be fow [full],
I'se no insist [on being included].

The second is little short of remarkable. We have only come to appreciate in the last decade that the core difference between humans and other animals is the fact that humans can stand back from the world as we experience it and ask how it might be in the future. Animals cannot, for their noses, as it were, are thrust so firmly up against the grindstone of experience that they can never wonder whether the world could have been other than it is or why the world has to be the way we find it – the two questions that make both science and literature possible. The last stanza of 'To a Mouse' says it all:

Still thou art blest, compared wi' me!
The present only toucheth thee:
But och! I backward cast my e'e [eye],
On prospects drear!
An' forward, tho' I canna see,
I guess and fear!

The mouse takes the world as it comes, but we can reflect on the past and anticipate the future, and spend hours in angst and fear because of it. I rest my case.

Latin in the dumps, science in decline

It has long been fashionable to decry the continued survival of Latin and Greek in the curriculum at some (usually rather posh) schools. It may seem odd to raise this in the context of a book on science, but as probably one of the few scientists around who can lay claim to an A Level in Latin, I feel I should rise to its defence.

I shall not dwell on the intrinsic interest of Latin as a language, nor on the window that its literature offers us on one of the most powerful and enduring cultures in the western world – despite the fact that its heritage colours much of our own language and a large proportion of our western European culture. Nor shall I comment on the fact that a significant proportion of the words we use have Latin roots, so that a knowledge of this supposedly 'dead' language can help us to understand the meanings of the words we use every day.

Instead, let me digress and begin with that eminent historian and raconteur, and sometime Fellow of Magdalen College, Oxford, A. J. P. Taylor. At a celebrated prize-giving one year at my rural grammar school, he caused near apoplexy among the staff (and not a few titters from the body of the hall) by advising us to ignore our lessons in favour of learning something *really* useful. And the most useful thing he had ever learned, he advised us in his inimitably avuncular way, was the complete list of all the sultans of Turkey.

Now, I never learned the list of the Turkish sultans, but I was, at the age of eight or nine, obliged to learn the rhyme for the kings and queens of England from 1066 onwards. For those of you who don't know it, it's very simple and here it is:

Willy, Willy, Harry, Ste*;
Harry, Dick, John, Harry Three;
One, two, three Neds, Richard Two;
Henries Four, Five, Six, then who?
Edwards Four, Five, Dick the Bad;
Harries twain and Ned the lad;
Mary, Bessie, James the vain;
Charlie, Charlie, James again;
William and Mary, Anna Gloria;
Four Georges, William and Victoria.

Now, apart from the fact that I have never been lost in discussions of the political history of England, its main contribution to my intellectual growth was, I am absolutely convinced, the training of my memory.

We all of us, in the final analysis, depend on our memories for a great deal of what we do. Sheer intuitive ingenuity is never enough for science to advance. Like any discipline, it depends on what in the humanities is sometimes referred to as scholarship – which is just a polite way of saying the ability to remember things. Advances in science, as in all forms of knowledge, come from being able to relate different events or things in new ways. Without the ability to remember the fine details of how the world actually is, no amount of intuition will allow even the proverbial genius down the hall to produce a new idea wholly independently of any remembered facts. Even mathematicians depend on memory to be able to

* Stephen.

recognise which of several possible ways of solving a mathematical problem is the most appropriate.

Recent developments in neuroanatomy seem relevant here. Current thinking on the development of the brain is coming round to the view that neurons initially lay down connections with each other at random and in immense numbers, but that these connections are whittled down by a process akin to natural selection during the first few years of childhood. Connections that are rarely used wither away and are lost; those that are regularly used are strengthened and increase in efficiency.

I hazard the guess that rote learning plays an important role in developing an individual's capacity to memorise and that much of this capacity is laid down at quite an early stage by this process of neural reinforcement. It is not for nothing, after all, that we teach nursery rhymes to children: their rhythmicity makes them particularly easy to learn and the story lines make them sufficiently interesting and fun to be worth the effort.

Which brings me back to Latin, for nowhere else was rote learning traditionally quite so important as in the morass of regular and irregular verbs of that language, and in the declensions and conjugations of its convoluted grammar. But what makes Latin different from both nursery rhymes and most other languages as a basis for training the mind is its great precision and its systematic structure (the very features that attracted bureaucrats to it long after Rome's decline). It provides a training not just in memorising, but in precisely those modes of thought that underpin everything we do as scientists. It is the perfect counterpoint to English, whose fluidity, lack of structure and enormous vocabulary are its very strengths as a literary language.

So my appeal is not to the hackneyed Victorian virtues of rote learning for the sake of parroting knowledge, but to the crucial role that rote learning seems to play in our intellectual development. In all our enthusiasm for new ways to make the school curriculum more interesting and more relevant – both laudable objectives in their own terms – we ought not to overlook the functions that apparent anachronisms in the curriculum actually served. Appearances are too often deceptive.

Chapter 18

Are You Lonesome Tonight?

In the Darwinian world of natural selection, reproduction is the motor of evolution. Success in the business of reproducing means making one's biological mark on the species' future gene pool, though it all depends on producing offspring that in their turn reproduce. Being a grandparent is what the evolutionary processes are all about. But producing offspring in either generation is only the end point of a long process that begins with courtship and choosing a good mate. Darwin hovers over our shoulders as we make our choices.

In traditional societies, men seek women who are young and fertile, while women seek men with prospects of status and wealth. Consider the marriage patterns of eighteenth- and nineteenth-century German peasants. Eckart Voland's researches into the parish registers of Krummhörn (see Chapter 4) showed that, matched for age, wealthier landed peasant farmers married significantly younger brides than landless day labourers did. In addition, it was clear that the women from the lower socio-economic classes were trying to hold out as long as possible for the opportunity to marry up the social scale.

For women, the benefits of marrying up the social scale

were significant. The wives of men higher in the social scale produced up to a third more surviving offspring, mainly as a result of higher rates of infant survival rather than higher birth rates. So the benefits of hypergamy (marrying up the social scale) were enormous. Not every woman could expect to succeed, of course. Eventually women of low status would be forced to cut their losses and make the best of a bad job within their own social circle. Like Jane Austen's eligible spinsters, they were eventually forced to bale out of the competition for Mr Darcy and settle for the curate when they felt that time was no longer on their side.

How to advertise and win friends

Lonely Hearts columns have come to be an important venue for contemporary mate-finding. So they provide us with a unique glimpse into the bargaining processes that underpin our choice of mate, a glimpse of what characteristics people seek in a partner and those they believe a prospective mate might be looking for in them. They amount to the opening bids in what in some cases will turn into a long chain of negotiation ending with some form of long-term relationship or marriage.

Devotees of Finlay MacDonald's wonderfully evocative account of his childhood in the Western Isles between the wars, *Crowdie and Cream*, will remember that Old Hector agonised a good deal about how to find himself a wife – not just about what the rest of the village might say if their aged bachelor turned up with one, but also how you went about finding someone suitable when living in a remote island community. The answer, as the worldly-wise eleven-year-old Finlay pointed out, was to advertise. Finlay's carefully constructed advertisement duly appeared in the *Stornoway Gazette*:

Retired seaman wants woman used to croft work with a view to matrumony [*sic*].

It had all the directness and lack of delicacy – as well as spelling mistakes – that an eleven-year-old could muster. But it worked. Hector was even spoiled for choice: he had three replies. Finlay's advice was to plump for the one that could spell best, commenting as an afterthought that she 'sounded like a good woman'. Whether by luck or instinct, he turned out to be right, and Hector lived into a contented old age with his Catriona.

Personal adverts have remained a popular means for finding love to this day. Think of it as the opening bid in a game of poker where, thanks to years of experience in the playground of life, you have some general rules about the kinds of things that appeal to the opposite sex, but no knowledge at all of who is actually out there looking for a mate. The name of the game is to stay in the frame – to ensure that you get enough replies that, like Old Hector, you can at least choose from what's on offer.

Most of us take the unwritten rules of this contractual bidding for granted. We accept that younger women find it easier to attract eligible men. We accept, too, that elderly male millionaires are more likely to marry twenty-year-old models than are their poorer contemporaries. But what are the origins of these preferences and to what extent do they influence our search for partners?

First the preferences. Psychologists Douglas Kenrick and

Richard Keefe of Arizona State University at Tempe have examined more than one thousand Lonely Hearts advertisements from the US, Holland and India. Their findings confirm what most of us might already suspect. As male lonely hearts age, they seek women who are increasingly younger than they are; they tend to opt consistently for women who are at the peak of fertility (in their late twenties). Female lonely hearts, by contrast, tend to prefer men who are three to five years older than themselves, with the age gap tending to diminish as they get older. So we end up with an inevitable mismatch: men want younger women, but women want men more their own age. In most cases, real life intervenes to find a compromise, since it's better to accept second best than have nothing at all. However, as the choosier sex, women have a slight advantage. What that means in practice is that they can afford to trade one trait against another with less disappointment because they have a greater number of traits to choose between. Older men only get young catches when they have something else to put on the table – and that invariably means wealth, and lots of it (or its surrogate, fame).

This is a particular problem for older women, because men's first thoughts focus so heavily on youth. Knowing that they have a weaker hand, older women are less demanding in their ads, seemingly being more willing to settle for anything rather than nothing. Catriona honestly declared her age, and offered nothing but her loneliness as a fifty-year-old spinster to entice Old Hector. But her sting in the tail was to call down the wrath of the Almighty if Hector was intent on making a fool of her. She was putting Hector to the test, while at the same time recognising that, in reality, her own choices were very limited.

Some older women get around this by not mentioning their age. This allows them to behave more like women in their twenties, in particular by being much more demanding than women who declare their age. More importantly, it allows them to stay in the game longer and at least retain the capacity to choose between respondents. The chink in the armour on this one is they still show the same age-related preference for a partner of similar age. So, if she doesn't say how old she is, just take five years off the age of the partner she is looking for and you won't go far wrong.

But age is only one criterion. What do the columns reveal about looks and money? To find out, David Waynforth, now at the University of East Anglia, and I analysed nearly nine hundred advertisements in four US newspapers. Male advertisers were more likely than females to seek a youthful mate (forty-two per cent of the men versus twenty-five per cent of the women) or a physically attractive one (forty-four per cent versus twenty-two per cent). No surprises there, perhaps. But male advertisers were also more coy about their own looks. We found that while fifty per cent of female lonely hearts used terms such as 'curvaceous', 'pretty', or 'gorgeous', only thirty-four per cent of the males used comparable terms ('handsome', 'hunk' or 'athletic').

It was a different story with money and status. Here, it was the female lonely hearts who made most demands. When specifying their requirements in a mate, they were four times more likely than males to use terms like 'college-educated', 'homeowner', and 'professional' as desirable in a prospective partner – all indicative of earning power or prospects. Male lonely hearts, on the other hand, were much keener than women to advertise such attributes. The cues can be quite subtle. In London, men will declare their postal area if it is up-market (Kensington or Hampstead), but never if it is down-market (Hackney or the Isle of Dogs).

Of course, no two cultures are the same, and the magnitude of these differences between the sexes is bound to vary from place to place. What surprised us, however, is how robust the general trends are. For example, when Sarah McGuinness and I studied six hundred ads placed in two London magazines, we found trends similar to those seen in the US ads. Sixty-eight per cent of women advertisers offered cues of physical attractiveness, compared to only fifty-one per cent of men.

There is a consistency, too, with findings from other types of research. One well-known scholar of the human 'mating game' is David Buss, a psychologist at the University of Texas, Austin. In 1989, he analysed questionnaires about marital preferences completed by over ten thousand people in thirty-seven different countries ranging from Australia to Zambia and from China to the US. Irrespective of culture, women tended to be more choosy than men, evaluating prospective partners on a much broader range of social and personality-based criteria. Women also consistently ranked the status and earning potential of a prospective mate higher than men did, while men rated youth and physical appearance more highly.

The mating game

The trends that we find in the Lonely Hearts adverts fit nicely with what we expect from evolutionary considerations. The biological processes of reproduction have very different implications for male and female behaviour, and so we would anticipate that men and women would focus on different aspects of the mating market. This is because, in mammals, the long-drawn-out processes of internal gestation and, later, lactation mean that males cannot contribute much in any direct sense to the business of reproduction once conception has taken place. This is a peculiarity of the fact that we are mammals. If human reproductive biology were more like that of birds or fishes, the story would be very different.

But mammals we are, so it is mammal biology that drives our mate-choice patterns. So males who want to maximise their reproductive success have only one option: to fertilise as many eggs as possible. For humans, that essentially means seeking a young, fertile partner with many child-bearing years ahead of her, or marrying as many women as possible at the same time. Females, on the other hand, are better placed to influence the infant's development directly. That means they are more likely to emphasise the business of rearing and look for mates with helpful resources. Wealth, status and occupation (all surrogates for wealth) feature highly as criteria in their ads. But they also give considerable weight to cues that signal commitment to the future relationship, and to cues that signal social skills. Men's ads then tend to offer those as self-descriptors – though you have to know how to read the code. Modern cues like 'GSOH' (good sense of humour) are intended to signal social skills, the ability to keep the partner interested and entertained.

The reason men place such a high premium on physical attractiveness in women? Once again, says biology, it is all to do with the quest for physical cues linked to age, health and, ultimately, fertility – cues that in the conventional world of our evolutionary past were difficult to fake. Take the case of women's typically hour-glass figure. Common experience suggests that men (by and large) prefer women with low waist-to-hip ratios, and research bears this out. Psychologist Devendra Singh, of the University of Texas, Austin, asked 195 men aged eighteen to eighty-five to rate drawings of women of different shapes and sizes from least to most attractive. The men rated women of average weight as more preferred than thin or fat women, but rated those with low waist-to-hip ratios the most attractive of all. Ratios of around 0.7 scored especially highly (healthy women in their twenties typically have waist-to-hip ratios of between 0.67 and 0.8). Significantly, perhaps, this turned out to be the shape of centrefold pin-ups from *Playboy* magazine over the past thirty years.

The preference is unlikely to be an accident of fashion. Women with low waist-to-hip ratios are on average more fertile than women with higher ratios. They enter puberty earlier and, according to studies of married women, conceive more easily. Although the precise reasons are not yet known, this almost certainly relates to the 'Frisch Effect', first identified by American reproductive biologist Rose Frisch in the 1980s: women only ovulate when their ratio of fat to total body mass reaches a certain level. The enlarged hips and thighs that give women their hour-glass shape are largely due to natural fat deposits in these areas. It seems that the wasp-waists and bustles of the Victorian period may have been attempts at exaggerating just these kinds of cues.

Similarly, our ideas about what characteristics go to

make a pretty face may also be rooted in the different reproductive strategies of the two sexes. Some of the most direct evidence has come from the neuropsychologist David Perrett and his laboratory at the University of St Andrews. Using composite pictures built up from 'preferred' faces, he and his colleagues were able to piece together the features that people find most attractive.

Women seem to find especially attractive in men those features that indicate sexual maturity, such as a strong jaw line and a prominent chin, as well as traits such as large eyes and a small nose. In women, it is large pupils and widely spaced eyes, high cheekbones, a small chin and upper lip and a generous mouth that most men find attractive. Many of these female traits are characteristic of children and could signal youth and hence higher fertility. Men are also attracted by soft glossy hair and by smooth shiny skin – two of the features that the cosmetics industry has latched on to. Both are the product of high oestrogen levels and are therefore difficult-to-mimic cues of youth and fertility.

What's more, people from different cultures and races tend to agree on what constitutes beauty. Michael Cunningham, a psychologist at the University of Louisville, Kentucky, asked people from different racial backgrounds to rate faces of different ethnic origin for attractiveness. There was striking cross-cultural agreement over what features constitute a pretty face. Essentially, they are child-like qualities in women and signs of maturity in men. David Perrett and his colleagues have carried out similar studies of facial attractiveness in European, Japanese and Zulu populations, with very similar results. Beauty may not be just in the eye of the beholder after all.

So imperfect a world

Most of us cannot aspire to the clear-eyed coquettishness of a Winona Ryder or the rugged handsomeness of a Richard Gere at the height of their careers. Worse still, we are only at the 'right' age for a brief period during a lifetime. So how should we ordinary mortals find our mates? Here, evolutionary theory suggests you should adjust your strategy to make the best of what may otherwise be a bad job. In other words, lower your expectations and settle for the bargain basement. It's pure Jane Austen.

This is exactly what happens in the Lonely Hearts columns. In our study of American ads, David Waynforth and I found that people adjust their bids in the light of their circumstances. Older women (who are less fertile) were less demanding in the traits they asked for in prospective mates than younger women were. Similarly, when matched for age, women who considered themselves physically attractive were more demanding than those who made no mention of appearance. If you think you have a strong bidding hand, you play the market for all it's worth.

The men in our Lonely Hearts study also modified their bids – not according to their looks but in the light of whether or not they offered cues of status or wealth. When matched for age, men who advertised cues of wealth and status were significantly more demanding of prospective partners than those who did not. Such men, for example, were less likely to tolerate children from a previous relationship. And unlike their female counterparts, male lonely hearts became more demanding of prospective mates as they aged, reflecting the growing strength of their hand in the poker game. The crunch point, however, came in middle age. Once past the mid-fifties, male advertisers lowered their demands, perhaps realising that mortality was making them an increasingly risky bet.

This kind of sensitivity to circumstances may even operate in relatively casual encounters between the sexes. James Pennebaker, a psychologist at the University of Virginia, asked (sober) men and women in singles bars to rate the other customers for attractiveness on a scale of one to ten. As closing time drew nearer, and hence the likelihood of heading home alone increased, they began to rate members of the opposite sex as increasingly more attractive. On average, members of the opposite sex were judged to be about twenty per cent more attractive at midnight than they were at 9 p.m. In contrast, their ratings of members of their own sex showed no such tendency to change with time. Since it seems at best implausible that the least pretty girls had been chosen first and gone off to their evening's tryst, it must be that the punters were progressively lowering their standards with respect to sexual partners as the prospect of failure loomed ever larger.

Children are a particular disadvantage to those seeking a new relationship later in life. Voland found that in the eighteenth- and nineteenth-century Krummhörn population young peasant widows who had had a single child from their first marriage had a seventeen per cent higher chance of remarrying if that child had died. We found an analogous trend in our Lonely Hearts sample from the US. Women who stated that they had young children from a previous relationship set their sights significantly lower than those who did not: when matched for age, women without dependants asked for almost twice as many traits in a prospective partner as women who had dependent children. Women with dependent children were forced to be a lot less choosy.

Life's little lessons

Most people are quite sensitive to the bargaining power they hold in the mating marketplace. In the early 1980s, psychologist Steve Duck, then at the University of Lancaster, ran a telling experiment in which male subjects were asked to complete a questionnaire for some (fictitious) research project. Present in the room at the same time was a young woman ostensibly engaged in the same task; in fact, she was a stooge who adopted different styles of dress and behaviour with different subjects. Duck found that the men's willingness to strike up a conversation with the stooge depended on the perceived similarities in their respective social styles. Yet again, it seems we pitch our bids to what we think we can get away with, and don't try to overbid the hands we hold. The mating game is unforgiving: what you can achieve in this game is not just of your choosing – it depends on someone else choosing you.

Boguslaw Pawlowski and I found a comparable sense of realism in UK Lonely Hearts ads. We calculated a simple index of selection for each sex – the ratio of the expressed preference for partners of a given age by members of the opposite sex relative to the number of individuals of that age in the advertising population. A selection ratio greater than 1 means you are in great demand; below 1, and you are less than popular. We then plotted how demanding advertisers were in relation to this selection ratio. For both sexes, the higher the selection ratio, the more demanding those individuals were in their search for partners. Except for one age group: men in their later forties. It seemed that they wildly overestimated the strength of their bargaining hand, and were far more demanding than their attractiveness to women actually warranted. Still, by their fifties they had learned the hard truth and radically downgraded their demands. So even men can learn, it seems.

This suggests a strong role for realism in the mating marketplace: no point investing resources trying to date someone too far above you in the social scale. We learn in the sandpit of life how we stand in the mating market, and adjust our aspirations accordingly. We might yearn for a Winona Ryder or a Richard Gere in our dreams, but it takes only a couple of cold shoulders for a sense of reality to intervene. That realism may partly explain why like tends, in the end, to settle for like when they make their final choice, their aspirations notwithstanding. Except in societies where arranged marriages are common, people are statistically more likely to marry those who are similar to themselves, not just in social and cultural background, but also in physical appearance. Among the more bizarre correlations between married couples, for example, is the relative length of the joints of the fingers.

Experience plays a particularly important influence in mate choice. This sensitivity to experience may explain one striking feature of our US Lonely Hearts sample, namely the frequency with which women advertisers sought traits linked to pairbonding and the family environment – traits signalled by words like 'loving', 'warm', 'GSOH' (good sense of humour), 'family-minded', 'gentle', 'dependable'. Some forty-five per cent of the women in our American sample desired at least one of these traits in a prospective mate, compared to only twenty-two per cent of men. Yet men did not advertise these traits any more often than women did, suggesting that men had not yet picked up on this change in women's concerns.

This probably reflects a cultural lag in the aspirations of the two sexes. It is quite clear that in traditional societies all over the world, wealth is the single most important factor influencing a woman's ability to rear offspring successfully. As a result, women place a very high premium on wealth (or at least future wealth potential) in their husbands. But the industrial revolution of the last century has had an important impact on women's ability to rear offspring in the industrialised West, in two crucial respects. First, dramatically improved medical technology has reduced childhood mortality to very low levels compared to what it was, and indeed still is, in pre-industrial societies. Second, the expanding economies of the industrialised countries have meant that wealth differentials are much less important in determining what you can afford to invest in child-rearing. In addition, women are now able to earn their own way and are no longer so dependent on their menfolk to provide them with the resources they need during the arduous and costly business of childcare.

With wealth per se no longer so important for women, the other (principally social) aspects of the rearing environment will have a much bigger impact on the success with which a woman rears her children. Hence the forty-five per cent of female lonely hearts who ask for 'caring, sharing' partners. But if women's priorities in the West have changed, the message from the Lonely Hearts data is that men have not yet realised this. Women might be seeking caring, sharing partners but men are still pushing the age-old traits of manliness and wealth for all they are worth.

Advertising is, of course, a shady business, and the business of mate searching is no different. Indeed, one of the commonest complaints made by people responding to Lonely Hearts advertisements is that the advertiser turned out to be nothing like their description. I suspect that most people actually have quite a realistic appreciation of their own worth in the mating marketplace and ask for traits in a partner that are a much better match to their real character than their descriptions of themselves (which tend to be overblown in order to keep their options as wide as possible).

So, if you're thinking of dipping into the Lonely Hearts columns, you might be advised to ignore what advertisers say about themselves and concentrate on what they ask for in a partner. It is probably a much better predictor of what they are really like. Otherwise, it's a game of poker.

Chapter 19

Eskimos Rub Noses

In July 1838, the young Darwin sat down and wrote out a list of pros and cons for marrying his cousin Emma Wedgwood (she of the famous pottery family). But it seems that he was really wasting his time. Whether or not she would accept him would depend more on things much more basely biological than what either of them thought of the advantages and disadvantages of marriage. It seems that, though the young Darwin was blissfully unaware of it, evolution has saddled us with a whole series of cheap chemical tricks that play a far more important role in our behaviour than most of us would like to think. Just when we thought our much-vaunted brains had allowed us to rise above base nature, base nature emerges from the shadows once again to slap our wrists and remind us of our past.

Take kissing, for example. Of course, monkeys and apes nuzzle and muzzle each other, especially when grooming. But all this serious mouthing stuff that we do – no other species does anything like it. Even though it is sometimes said not to be universal in all human cultures, it's certainly very widespread and it isn't just a consequence of how close you've been to the French. So what's it all about?

Ae fond kiss?

Freud and his fellow travellers insisted that kissing was just some kind of reversion to infancy and the deeply buried memory of the pleasures of sucking on your mother's breast. Well, it's easy enough to see that adult kissing might have originated thus, but sucking breasts and kissing aren't quite the same thing. After all, if it really was a reversion to breast-sucking, why not just do *that*? Another suggestion is that it's a form of courtship feeding, a habit widespread in the insects and some birds. But that tends to be a male thing, with males offering packets of food (sometimes regurgitated, sometimes not) as gifts to prospective mates. Females evaluate the quality of a male by the size of the offering. It has a certain logic, rather similar to that of offering large diamond rings and mink coats to one's beloved. But it doesn't quite make sense when there is no food involved. And, in any case, we already do exactly this by other perfectly good means – think box of chocolates, flowers even. Besides, both sexes do kissing with equal enthusiasm, and courtship feeding is usually one-way. Something else is clearly afoot.

In fact, kissing is probably all about testing the genetic make-up of prospective mates. Our immune system is what defines us individually, and it is determined mainly by a little cluster of genes known as the major histocompatibility complex, or MHC for short. The MHC genes determine the range of foreign bodies (everything from pollen to viruses and bacteria) that your body can recognise and get rid of should they invade. It is a set of genes that is particularly prone to generating mutations, thereby allowing us to adapt to the threats posed by the ever-mutating microscopic world that continually parasitises us and threatens our personal survival. The MHC genes also control your odour, because it turns out that your natural smell is closely related to your immune response.

There has been a long series of studies which have shown that people tend to prefer to mate with people who have complementary MHC genes. The reason is fairly obvious. If you mate with someone with the identical immune response, your children will only have that limited immunity. But if you mate with someone with a complementary set of responses, your children will have a much wider range of immunity to the diseases that threaten them.

So how do you get to find out whether a prospective partner has the right set of immune responses to suit you? Smell is one way of doing it, and smell obviously means getting up close and personal. That's why our perfume preferences are very personal: they seem to be directly related to the natural body odour that we have. In effect, we prefer to wear those perfumes that enhance our natural smell – that's why it's always tricky buying perfumes for someone you don't know very well. But smells can be masked, not just by ladling on Givenchy's latest, but also, in the state of nature in which we have spent most of our evolutionary history, by accumulations of dirt and bacteria. So one way to circumvent this problem is to get up even more close and personal and taste the stuff directly.

Saliva is full of chemicals exuded by the body, not the least of which are a group of proteins known as MUPs (major urinary proteins). OK, this doesn't sound so good. But before you panic, the name comes from the fact that MUPs were first identified in rodent urine, where they seem to have a lot to do with individual recognition and territorial behaviour. Jane Hurst and her colleagues at Liverpool University have recently shown that female mice can discriminate between males solely on the basis of differences in their MUPs. MUPs occur in urine simply because urine is a very convenient mechanism for animals to deposit signals of their presence in an area. They also probably occur pretty much everywhere that you care to excrete bodily fluids of any kind.

So, the next time you get in a deep clinch, you might pause and remind yourself that this is all about choosing the right mate with a good set of immune responses to complement your own and that MUPs are the route to success . . . On second thoughts, perhaps you should just shut down your busy conscious mind and let your subconscious take over so that nature can take its course. Evolution did not spend many millions of years perfecting a mechanism of mate choice only to have you mess it up by thinking too much about it.

Eskimos rub noses

Now, if there is one thing everyone knows about Eskimos, it's that, instead of shaking hands, they rub noses when they greet each other. Well, actually, that's a bit of a myth created by European explorers when they first came across the Eskimos. In fact, they place their noses next to each other's faces and breathe in deeply. Nor are they the only people to do this. The Maoris in New Zealand also rub noses when they meet, a behaviour known as *hongi*. Theirs too is less of a rub and more of a light press of one nose upon another in a symbolic joining together of host and visitor.

What these folks are in fact doing is breathing in each other's smell, one of the best markers of who you really are. In our vision-dominated world, we often forget just how important smell is to us. In fact, we use smell a great deal more than we realise – and nowhere more so than in the business of mate choice. Back in the mischievous 1960s, some experimenters sprayed androstenone (one of a family of steroids that are a natural by-product of testosterone, the so-called male hormone – it's responsible for the slightly musty smell that aftershave-less men often have) around some of the cubicles in men's and women's public toilets. Then they sat back and watched. What they found was that men avoided the 'androstenoned' cubicles – having ventured in, they would usually back out hastily and find an androstenone-free one instead. But women made a bee-line for the androstenoned cubicles.

In an updated version of that experiment, Tamsin Saxton and her colleagues at Liverpool University applied androstadienone (another of the same family of steroids) to the upper lips of women at a speed-dating event. In a speed-dating event (for those of you who have yet to experience this curious form of mating market for the ultra-busy), the girls sit round the room at tables and the boys spend five minutes in a brief conversation with each one in turn, in a kind of gigantic round-robin. At the end of the evening, each person lists the names of the people they would like to meet again, and the organisers then exchange details for those who express an interest in meeting each other. It's a perfect setting for an experiment in which each sex can have a brief taste of a dozen or so possible mates and hopefully choose the most congenial.

In this study, the androstadienone was concealed in

clove oil to disguise it, which also made it possible to control for the effects of other odours. So a third of the women had androstadienone-plus-clove-oil, another third had just clove oil and the final third had plain water. That way, it was possible to separate out clearly the effect of the clove oil substrate from the smell itself.

The results were spectacular. The women who had received the androstadienone not only rated the men they met at the speed-dating event as more attractive than the women in the other two groups, but they were also significantly more likely to ask to see them again. Somehow, the androstadienone acts on a deeply buried brain mechanism to precipitate a rosier-than-reality view of the hulking brute that stands before you. Who said romance was dead?

Who dares, wins

Still, when all else fails, guys, there is one way to improve your chances. Become a hero. Some years ago, Sue Kelly, then one of my students, ran an experiment in which she offered women a series of vignettes of men with a range of different traits. Some were boringly steady with humdrum jobs, some worked in caring professions, others were risk-takers. The women were asked to rate each individual for their attractiveness as a friend, a long-term partner or a prospect for a one-night stand. The caring altruists got top marks as long-term partners, but the risk-takers swept the board as partners for one-night stands. They were simply rated as much more attractive. William Farthing of the University of Maine obtained similar results when he asked women to rate different males for attractiveness as mates: they much preferred heroic risk-takers against non-heroic risk-takers, although in both cases they rated those who took medium risks higher than those who took high risks. It seems that, in general, risk-taking is good advertising copy if you are a male, but don't overdo it: unnecessary stupidity carries a premium.

So do men take more risks than women? The answer, in general, is yes. And we found evidence of this when we carried out a study at a zebra crossing at a busy city-centre junction. In general, men took higher risks than women – in other words, they were more likely to cross the road when a car was approaching the crossing and the lights were on green for the car than women were. More importantly in the present context, they were much more likely to do this if there were women present as an audience than if there were not.

Is this because men realise that women are attracted by risk-taking behaviour? The answer seems to be that men are quite good at identifying the things that press women's mate-choice buttons. In her study, Sue Kelly wanted to know whether men understood women's preferences, and so asked men to rate the same character vignettes from a woman's point of view. They were pretty good, though they tended to exaggerate women's actual preferences.

Several recent studies have looked at real-life heroism from an evolutionary point of view. One study examined the records for the Carnegie Medal, a very prestigious national award in the USA given to civilians for unusual courage in emergency situations – for example, rushing into a raging torrent to save someone's life. The citations for these awards revealed some very striking patterns. Men were much more likely to rescue (or attempt to rescue) unrelated young women than anyone else, whereas women were disproportionately more likely to rescue related children. In other words, for women acts of heroism are about investment in one's children, but for men they appear to be more about mating opportunities. Another of my students, Minna Lyons, analysed a large corpus of recent British newspaper accounts where people had attempted to rescue others in distress. Almost all the rescuers were men, but there was an intriguing status bias. Men from the richer end of society rarely acted as heroes; instead, most of the rescuers were from the poorer end of the socio-economic spectrum. Such men, she argued, had more to gain in the mating market from being recognised as heroes.

I found a rather analogous trend among the historical Cheyenne Indians of North America. The Cheyenne had two kinds of chiefs: peace chiefs who inherited their status, never took part in war and married early, and war chiefs who eschewed marriage and led the tribe in war, often staking themselves to the field of battle so as to die rather than be defeated. A war chief might eventually marry, but only if he survived long enough to be able to renounce his war vows with honour. The demographic records from the later nineteenth century show that men from the hereditary class of peace chiefs (the upper tier of society) almost never became war chiefs. Instead, almost all the war chiefs were orphans or the sons of low-born members of the tribe, whose chances of finding a wife were slim at the best because their status made them less than desirable catches. But men who were successful war chiefs – meaning those who survived long enough to be able to retire with honour and rejoin normal society –

proved to be very attractive. On average, they ended up siring more children than peace chiefs despite having a much shorter married life.

That risk-takers are reproductively more successful seems still to be true, even in the more peaceful environment of modern Britain. Giselle Partridge, then one of my students at Liverpool University, carried out an extensive survey of men's risk-taking and compared this with the number of children they had during their lifetime. She measured risk-taking both through occupation (firemen, for example, compared to desk-bound administrators) and through a questionnaire on behaviour (convictions for speeding, risky leisure activities). High risk-takers had significantly more children than low risk-takers. While the explanation remains unclear (are high risk-takers more likely to have unprotected sex, or are they just more attractive to women?), the facts speak clearly enough. Men that take risks make a bigger contribution to the next generation.

You pays your money, and you takes your choice.

Chapter 20

Your Cheating Heart

Some years ago, my former colleague Sandy Harcourt, now at the University of California at Davis, showed that primates which mated monogamously had much smaller testes for body weight than species that mated promiscuously. To evolutionary biologists, the explanation was obvious. In promiscuous mating systems, males can never be sure that they will be the one who is mating with a female at the time she ovulates and an egg is available for fertilising. In such cases, the best way to maximise the chances of fertilising the female is to leave her with as much sperm as you can possibly muster in order to swamp the sperm of any other males that have previously mated with her, or might mate with her in the next few days during her fertilisation window. To achieve that, it is necessary to have large testes capable of producing excess quantities of sperm. The great quandary for us is that when Harcourt and his colleagues plotted humans on the graph, they fell exactly halfway between the two groups – we are neither wholly monogamous nor wholly promiscuous. So are we monogamous or promiscuous by nature?

'Til death us do part

With these words, Christianity has traditionally enshrined the idea that humans are a monogamous species. So why do more than a third of all marriages in Britain and half of those in the US end in divorce? And how come as many as fifteen per cent of children are not the biological offspring of their registered father? Some people see this as a sign of the times: a breakdown in family values, the disintegration of society, or a modern disease that requires everything, including relationships, to be 'new and improved'. In recent years biologists have come up with another explanation. They have been finding that monogamy is not a fixed and immutable instinct, hard-wired into an animal's brain. Even creatures once considered as paragons of fidelity will indulge in a fling, if the situation is right.

Take the South American marmoset and tamarin monkeys. Both are usually monogamous in the wild, with males largely responsible for bringing up the young. But in some cases, males engage in roving polygamy, hitching up with a succession of females. 'Divorce' rates can be as high as a quarter to a third of all pairs in the population during any one year. This radical change in behaviour is often prompted by an excess of males, usually because of high female mortality. With females in short supply, males who cannot get a mate become 'helpers-at-the-nest', willing to assist with the rearing of offspring that are not their own. The presence of a helper increases the chances that a breeding male will desert his partner and go searching for another female, because he will be able to breed again sooner than if he waits for his current mate to come back into breeding condition. The helper gets his payoff next time the female comes into oestrus, when he has his turn to mate with her. And the females seem indifferent to their mate's behaviour: so long as they have a male to help with rearing the offspring, they don't seem to mind too much who that male is.

Breeding males who are powerful enough to pursue this kind of roving strategy can gain up to twice as many offspring as they would by remaining in a normal monogamous relationship. Females fare no better or worse, while the helpers make the best of a bad situation. In other words, flexible behaviour patterns allow the breeding males to exploit the shortage of females to increase their reproductive success. In this case, the new behaviour is a response to a change in circumstances. But even without external changes, it may be in the interests of monogamists to adopt a more flexible approach. The animal world, it turns out, is full of examples of cuckoldry, cheating and even divorce, by supposedly lifelong mates as they try to overcome what I call the monogamist's dilemma.

Monogamy is relatively rare among mammals. Only about five per cent of mammals are monogamous, with primates and the dog family (wolves, jackals, foxes etc) favouring the practice more than most. But there is one group of animals for which monogamy is the rule. Around ninety per cent of bird species pair, at least for a given breeding season. On the surface, it looks like true wedded bliss. But a decade or so ago, that illusion was blown away when the new technology of DNA fingerprinting revealed that as many as a fifth of the eggs produced by supposedly monogamous female birds had not been sired by their regular partners. Many male birds were busily feeding offspring that were not their own.

What on earth is going on? Behavioural ecologists, who had previously focused on co-operation as the driving force behind monogamy, had to revise their views about mating strategies. They began to see the flip side of the coin: that together with co-operation comes the inevitable risk of exploitation. Monogamous males can never be sure that they are the father of their partner's young. In all co-operative systems, it always pays some individuals to opt for the free-rider strategy by leaving their friends – in this case, literally – holding the baby. That way, they gain all the benefits without having to pay the costs. The monogamist's dilemma is whether to stay with your mate and risk being cuckolded, or to abandon family life and risk losing the offspring you have just sired because their mother cannot rear them on her own.

Males would like to have it all. And in evolutionary terms that means developing sneaky strategies to mate with new females, while finding ways to avoid wasting energy bringing up the offspring of other males. Once DNA analysis revealed the extent of extra-pair mating, researchers began to see the mating game for what it was and anti-cuckoldry strategies also started to be noticed. It also works the other way round, of course. Perhaps the best-known example is provided by the humble dunnock, the small and undistinguished-looking British hedge sparrow. Nick Davies and his colleagues at Cambridge University have shown that male dunnocks adjust the effort they put into bringing food back to the nest exactly in proportion to the number of nestlings they have sired (as determined by DNA fingerprinting). And how do they achieve this remarkable feat? By the very simple trick of estimating how much time the female was out of their sight during the egg-laying period. That, it turns out, is a very good estimate of the likelihood that she was engaging in a little fling in the bushes with the chap next door.

Humans are also highly suspicious of extra-marital relationships, a fact attested to by the frequency with which separated husbands are now resorting to DNA fingerprinting to avoid paying their former wives for the upkeep of children who aren't in fact theirs. And it seems they may be justified. A few years ago, Robin Baker and Mark Bellis, then both at the University of Manchester, calculated that between ten and thirteen per cent of all conceptions in the UK arose from matings with non-pair men. They based their estimate on self-reports of the frequency of double matings – matings with both the normal partner and another man within five days of each other – at around the time of ovulation.

In some cultures, males attempt to sequester their wives in harems where the opportunities for infidelity are greatly reduced, or persuade them to wear blandly uniform clothing supposedly for religious reasons. Such behaviour is just a form of mate-guarding, no different to the numerous examples seen in many animal species. Where society favours less formally structured relationships, men and women seem, at a subconscious level at least, to realise that paternity can be an issue. That's one reason why, as I mentioned in Chapter 8, in-laws make so much fuss about newborn babies looking like their father. It looks suspiciously like an attempt to persuade the husband that the baby really is his and so encourage him to invest in the baby.

However, careful analyses of the evolutionary costs and

benefits of rearing another man's offspring suggest that a male's response to suspicions of cuckoldry should not necessarily be outrage. Although a male risks rearing children unrelated to him, he would do best in the long run if he treated all his partner's children as his own, as long as doing so allows him to maintain a satisfactory relationship with her and thereby gain access to most of her future reproduction. Being too inquisitive may backfire by raising too many doubts in his own mind or by causing his partner to desert him in favour of a kinder rival. Rearing a few offspring sired by another male may simply be the cost some males are obliged to bear in order to reproduce at all. It seems that Freud might have underestimated the benefits of repression.

Monogamy on the rocks

It is easy to see what a monogamous male gets from playing away from home. But it takes two to tango, so what does the female gain from acquiescing in an extra-pair relationship? Current evolutionary thinking on this emphasises two possibilities. The first can be described as bet-hedging. Ideally the female would like a male who will invest in her offspring: a man with a bulging wallet, perhaps, or a robin with a large breeding territory. But she also wants a mate with good genes, something she might assess by looking at his tail if she is a peahen, or by the symmetry of his features if she is a woman. But females usually have to trade one component off against another because the world is imperfect and few males come with high ratings on all dimensions – and those that do are usually swamped by suitors. So perhaps she could try to get the best of both worlds by teaming up with a good provider and allowing him most – but not all – of her conceptions, while allocating the rest to better-quality mates as and when she can.

An alternative explanation for females' interest in extra-pair matings is that it is a way of forcing their pair-male to be more attentive. Magnus Enquist and his colleagues at Stockholm University have used a simple mathematical model to show that females can play one male off against another in this way to prevent their pair-male straying in search of other females with whom to mate. But, once again, there is a fine line to tread. Martin Daly and Margo Wilson have shown, using data from all around the world, that the vast majority of spousal murders in humans are triggered by actual or suspected infidelity. Both men and women often use aggression as coercion to try to prevent a mate abandoning them, but sometimes males overplay their hand too heavily.

Even so, intra-sexual jealousy seems to be the first line of defence for maintaining the pair-bond in many species. In titis, one of the many small monogamous South American monkeys, females are quite intolerant of the approach of strange females, and will drive them away. And I have observed similar behaviour during fieldwork on the small monogamous African antelope known as the klipspringer.

Maria Sandell of Lund University, Sweden, studied this experimentally in European starlings. During the egg-laying period, stranger females were placed in small cages near to the nest-box used by an established wild pair. Males offered the opportunity of a second female showed considerable interest, but their females were rather aggressive towards the rival. More importantly, Sandell was able to show that females who were more aggressive towards the rival were more likely to retain a monogamous relationship with their male throughout the breeding season than were less aggressive females.

Nonetheless, evolutionary interests suggest that individuals should be open to better reproductive opportunities that happen to come their way. So we should not be surprised to see partnerships being dissolved as new and better opportunities come along. Researchers are finding that 'divorce' is common even among birds like swans that supposedly pair for life. Estimates of pair-bond dissolution vary enormously, both across species and, within species, across populations. André Dhondt, now at Cornell University, found that over half of all pairs of Belgian great tits, for example, get divorced. Not only did the females often instigate divorce, but they usually benefited by subsequently producing more offspring when they did. The males, however, did not always fare so well.

Failure to rear offspring is one common cause of avian divorce. Failure to have children is also one of the highest risk factors for divorce in humans, and not just among Muslims. (Under Islamic law, a wife's infertility is an appropriate reason for divorcing her and sending her back to her parents. Infidelity by a wife – but not by a husband – may even be punishable by death.) However, there are as many other routes to divorce in avian society as there are among humans. Lewis Oring of the University of Nevada, Reno, studied killdeer, a North American plover, and observed 'home-wreckers' – individuals that muscle in on another pair and drive the same-sex member out so that they can take over its mate. Bob Furness of the University of Glasgow has seen similar behaviour in great skuas, a sea bird whose ferocious reputation is richly attested to by the fact that its attempts to oust a member of an established pair may sometimes result in the luckless victim's death.

If there is a message in all this, it must surely be that there are no simple rules that apply to all species all of the time. As is invariably the case in biology, there are some key general principles that apply universally, but the patterns of monogamy, divorce and polygamy vary both between and within species in response to the way these principles work themselves out in the local ecological and demographic conditions. Any animal with a decent-sized brain – and that most obviously includes humans – has its brain in order to tweak its behaviour to take advantage of the momentary circumstances it happens to find itself in. It is the availability of alternatives that makes shifts of behavioural strategy possible. Animals, every bit as much as humans, make choices about whom to pair with and how long for, and those decisions are influenced in large part by whether they will do better by staying with the current partner, by moving from one partner to another or by playing a more subtle kind of game.

Humans are caught in the same bind as any other monogamous species. The male wants to monopolise his mate's future reproductive output, but he has to tread a careful line. Mating is ultimately a game of co-operation not coercion: too aggressive a policing strategy may well put the female off and drive her away. In Californian chuckwalla lizards, for example, very aggressive territorial males achieve fewer matings because they scare females away from their territories. And Barbara Smuts, of the University of Michigan, has shown that overly aggressive male baboons suffer the same fate: females spurn their attentions in favour of socially more skilful males.

Just check out his DNA, my dear

A great deal of fuss has been made in the media about oxytocin, the so-called 'love hormone' that characterises monogamous species. In fact, oxytocin seems to have this effect only in females. In males, a related but rather different neuro-endocrine called vasopressin seems to be the active ingredient in these monogamous species. Vasopressin seems to play an important role in modulating male behaviour in monogamous species. When inserted into their brains, it makes male rodents more tolerant of females and young, more willing to engage in huddling behaviour and less aggressive. Inevitably, people have begun to wonder if it plays a similar role in humans. Given the difficulty we have in deciding whether humans are monogamous or promiscuous, maybe the issue is not so much that all human males should be high on the vasopressin dimension (and hence monogamous), but rather that there are differences among males that might correlate with promiscuous behaviour.

Hasse Walum, of the Karolinska Institute in Stockholm, and his colleagues used a large sample of 552 Swedish twins to look at the relationship between vasopressin receptor genes and marital stability in men. They checked out a number of genes in the region that codes for the vasopressin receptor. They found that one particular gene site, RS3, varied significantly as a function of the man's score on a partner-bonding scale that measured their commitment to relationships. And of the eleven different gene variants that occurred at this site, one in particular (allele 334) showed much the strongest effect.

Men who had one or two copies of the 334 allele (in other words, a copy inherited from one or both parents) scored lower on the partner-bonding scale than men who had two copies of any of the other ten alleles. They were also more likely to be living with, rather than married to, their partner – something suggestive of reduced commitment. One-third (thirty-three per cent) of double-334 men reported that they had experienced marital stress in the past year, against just sixteen per cent of single-copy 334 men and fifteen per cent of males who lacked the 334 allele. And all this despite the fact that all the men in the sample had been living with their partner in a stable relationship for at least five years and had at least one child with them.

In the Swedish sample, about four per cent of the men had two copies of the 334 allele and thirty-six per cent had one copy, leaving almost two-thirds of men having no copies and being a good bet for a devoted, monogamous partner. So, although the number of complete bastards (those with a double dose of the offending gene) seems to be very small, around a third of men seem to be a risky bet. A similar ratio was found in a very large survey carried out in Quebec by Daniel Pérusse: he found that around one third of the men in Quebec were habitually promiscuous, with about two-thirds being habitually monogamous (at least while they were in a steady relationship).

In the Quebec case, I was able to show that while promiscuous men would, as individuals, have sired more offspring over a lifetime than monogamous men (based on the frequency of copulation and the probability of conception occurring on any given copulation), the relative difference in siring rates between the two types of men exactly balanced their respective frequencies in the population. This suggests that monogamy versus promiscuity is a balanced evolutionary polymorphism, with the proportions of the two strategies held in approximate balance across the generations by the costs and benefits of pursuing the different strategies.

Although it's tempting to interpret these findings in terms of vasopressin being a male 'gene for monogamy', it almost certainly isn't – not least because the genetics of life are rarely so simple. Behaviour is often the outcome of predispositions laid down by the genes, rather than an outcome of the genes themselves. So it was interesting to see, in a recent study carried out by Dominic Johnson (now at the University of Edinburgh) and his colleagues, that males who had the RS3 gene were inclined to react aggressively when put in a threatening situation. It seemed that the RS3 gene simply causes men to fly off the handle quicker in response to something as simple as frustration. So men with the 334 allele are not genetically promiscuous: rather, they just don't think before they act.

So, girls, all of this seems to suggest that there is about a six in ten chance of picking a reliable partner if you choose at random from the population. Which rather makes it look like that old cigarette-butt trick might be the clever way to select a mate. Offer him a cigarette and, after he has finished, whisk the butt down to the genetics lab, where they can now squeeze out a sample of his DNA from the saliva stains and scan it for allele 334 at the RS3 gene locus. Not so good if it comes up positive. Very bad if it comes up double positive.

Chapter 21

Morality on the Brain

In 1906, New York's Bronx Zoo exhibited an African Pygmy in a cage next door to its gorillas, a spectacle that attracted huge crowds. Sadly, Ota Benga, the Pygmy in question, committed suicide in Virginia a couple of years later after being released, unable to cope either with the life he now faced in America or with the fact that he was completely cut off from his home in the Congo by what was in his impecunious circumstances an all but impassable sea journey. Today, we would regard this whole episode as an unacceptable breach of civil rights, an example of thoughtless cruelty and racism.

Our modern willingness to extend equal rights regardless of race reflects the belief that we are all of the same 'kind'. And we believe that to be the case because all of us, regardless of race, seem to share certain traits (notably the capacity to be moral) that make us all human. But how is it that we accord these rights to others? What makes us think we should? And where should we draw the line? These are thorny issues that have troubled philosophers for centuries, but might now be possible to answer thanks to insights from neuroscience.

Morality on the brain

That great paragon of the eighteenth-century Edinburgh Enlightenment, David Hume, argued that morality is mainly a matter of emotion: our gut instincts, the great man opined, drive our decisions about how we and others ought to behave. Sympathy and empathy play a significant role. But his equally great German contemporary, Immanuel Kant, took exception to what he saw as an entirely unsatisfactory way to organise one's life: our moral sentiments, he argued with equal insistence, are the product of rational thought as we evaluate the pros and cons of alternative actions.

Kant's rationalist view gained ascendancy in the nineteenth century, mainly thanks to the utilitarian theories of Jeremy Bentham and John Stuart Mill, who argued that the right thing to do was whatever yielded the greatest good for the most people – the view that underpins much modern law-making. Successive generations of philosophers have continued to argue the merits of both views.

However, recent advances in neuropsychology look like they are about to come down firmly in favour of good Scottish common sense. One such insight into how we make moral judgements has come from an elegantly simple series of experiments by Jonathan Haidt and his colleagues at the University of Virginia. They asked subjects to make judgements about morally dubious behaviour, but some did so while rather closer than they might have wished to a smelly toilet or a messy desk, and others did so in a more salubrious environment. The first group gave much harsher judgements than the second, suggesting that their judgements were affected by their emotional state.

One of the classic dilemmas used in studies of morality is known as the 'trolley problem'. It goes like this. Imagine you are the driver of a railway trolley approaching a set of points. You realise that your route takes you down a line where five men are working on the railway unaware of your approach. But there is a switch you can pull that would throw the points and send you off down the other line where just one man is working. Would you pull the switch? Most people would say yes, on the grounds that one certain death is better than five, and this is the Kantian rational answer predicated on the utilitarian view that our actions would maximise the greatest good.

But now suppose you are not driving the trolley, but standing on a bridge above the railway. Beside you is a giant of a man, of a size capable of stopping the trolley dead if you threw him off the bridge onto the railway line, so saving the five workers at the expense of this one luckless victim. Most people now hesitate to act so as to save the five workers, even though the utilitarian value is exactly the same – one man dies to save five. In most such cases, subjects cannot say why they have changed their minds, but one difference seems to lie in a subtle distinction between accidents and intentions.

The important role of intentions was borne out by a study of stroke patients, which showed that people with damage to the brain's frontal lobe will usually opt for the rational utilitarian option and throw their companion off the bridge. The frontal lobes provide one area in the brain where we evaluate intentional behaviour. The importance of intentionality has recently been confirmed by Marc Hauser from Harvard and Rebecca Saxe from MIT: they found that, when subjects are processing moral dilemmas like the trolley problem, the areas in the brain that are especially involved in evaluating intentionality (such as the right temporal-parietal junction just behind your right ear) are particularly active. Our appreciation of intentions is crucially wrapped up with our ability to empathise with others.

The final piece in the jigsaw has now been added by Ming Hsu and colleagues at the California Institute of Technology in Pasadena. In a recent neuroimaging study, they forced subjects to consider a trade-off between equity (an emotional response to perceived unfairness) and efficiency in a moral dilemma about delivering food to starving children in Uganda. They found that when decisions were based on efficiency, there was more neural activity in the areas of the brain associated with reward (particularly the region known as the putamen), whereas when decisions were more influenced by perceived inequality, it was areas associated with emotional responses to norm violations (such as the insula) that were more active. More importantly, the stronger the neural response in each of these areas, the more likely the appropriate behavioural response by the subject. In other words, judgements about morality and those about utilitarian efficiency are made in separate places in the brain, and may not necessarily be called on at the same time.

It seems that Hume was right all along.

A very peculiar species of morality

However, if morality is simply a reflection of empathy (and/or sympathy), then it seems unlikely that we really need a great deal more than second-order intentionality: it is only necessary that *I understand that you feel something* (or that *you believe something to be the case*). But morality based on this as a founding principle will always be unstable: it is susceptible to the risk that you and I differ in what we consider to be acceptable behaviour. I may think there is nothing wrong with stealing and be unable to empathise with your distraught feelings on finding that I have robbed you of your most treasured possessions. It's not that I don't recognise that you are distraught (or understand what it means for me to feel the same way), it's just that I happen to believe that theft is perfectly OK and that you're making a big fuss about nothing. If you want to steal from me, that's just fine . . . help yourself. I will surely try to defend my possessions, but my view of the world is that possession is nine-tenths of the law, and may the best man win.

If we want morality to stick, we have to have some higher force to justify it. The arm of the civil law will do just fine as a mechanism for enforcing the collective will. But equally, so will a higher moral principle – in other words, belief in a sacrosanct philosophical principle or a belief in a higher religious authority (such as God). The latter is particularly interesting because, if we unpack its cognitive structure, it seems likely to be very demanding of our intentionality abilities. For a religious system to have any kind of force, I have to *believe* that you *suppose* that there is a higher being who *understands* that you and

I *wish* something will happen (such as the divinity's intervention on our behalf). It seems that we need at least the fourth order to make the system fly. And that probably means that someone with fifth-order abilities is needed to think through all the ramifications to set the thing up in the first place. In other words, religion (and hence moral *systems* as we understand them) is dependent on social cognitive abilities that lie at the very limits of what humans can naturally manage.

The significance of this becomes apparent if we go back to the differences in social cognition between monkeys, apes and humans and relate these to the neuroanatomical differences between us. While humans can achieve fifth-order intentionality, and apes can just about manage level two, everyone is agreed that monkeys are stuck very firmly at level one (they cannot imagine that the world could ever be different from how they actually experience it). They could never imagine, for example, that there might be a parallel world peopled by gods and spirits whom we don't actually see but who know how we feel and can interfere in our world.

At this point, an important bit of the neuroanatomical jigsaw comes into play. If you plot the volume of the striate cortex (the primary visual area in the brain) against the rest of the neocortex for all primates (including humans), you find that the relationship between these two components is not linear: it begins to tail off at about the brain size of great apes. Great apes and humans have less striate cortex than you might expect for their brain size. This may be because, after a certain point, adding more visual cortex doesn't necessarily add significantly to the first layer of visual processing (which mostly deals with pattern recognition). Instead, as brain volume (or at least neocortex volume) continues to expand, more neurons become available for those areas anterior to the striate cortex (i.e. those areas that are involved in attaching *meaning* to the patterns picked out in the earlier stages of visual processing). An important part of that is, of course, the high-level executive functions associated with the frontal lobes. Since the brain has, in effect, evolved from back to front (i.e. the increase in brain size during primate evolution is disproportionately associated with expansion of the frontal and temporal lobes), it is precisely those areas associated with advanced social cognitive functions that become disproportionately available once primate brain size passes beyond the size of great apes. Indeed, great ape brain size seems to lie on a critical neuroanatomical threshold in this respect: it marks the point where nonstriate cortex (and especially frontal cortex) starts to become disproportionately available.

It seems to me no accident that this is precisely the point at which advanced social cognition (i.e. theory of mind) is first seen in nonhuman animals. Moreover, if we plot the achieved levels of intentionality for monkeys, apes and humans against frontal-lobe volume, we get a completely straight line. That, too, seems to me no accident.

So, we seem to have arrived at a point where we can begin to understand why humans – and only humans – are capable of making moral judgements. The essence of the argument is that the dramatic increase in neocortex size that we see in modern humans reflects the need to evolve much larger groups than are characteristic of other primates (either to cope with higher levels of predation or to facilitate a more nomadic lifestyle). After a certain point, however, the computing power that a large neocortex brought to bear on processing and manipulating information about the (mainly social) world passed through a critical threshold that allowed the individual to reflect back on its own mind. As we saw in an earlier chapter, great apes probably lie just at that critical threshold. With more computing power still, this process could become truly reflexive, allowing an individual to work recursively through layers of relationships at either the dyadic level (*I believe that you intend that I should suppose that you want to do something . . .*) or between individuals (*I believe that you intend that James thinks that Andrew wants . . .*). At that point, and only at that point, can religion and its associated moral systems come into being. In terms of frontal-lobe volume expansion, the evidence from the human fossil record suggests that this point is likely to have been quite late in human history. It is almost certainly associated with the appearance of archaic humans around half a million years ago. I'll come back to this in the next chapter. Before I do, however, let's explore the possibility of morality in other species a little bit more.

Can apes be moral?

Our nearest living relatives are, beyond any question, the great apes. Until only twenty years ago, it was widely accepted that the ape lineage consisted of two groups: modern humans and their ancestors on the one hand and, on the other, the four species of great apes (two chimps, the gorilla and the orang utan) and their ancestors. However, modern genetic evidence has shown that this classification, based largely on body form, is in fact incorrect. There are indeed two groups, but the two groups are made up of the African apes (humans, two chimpanzees and the gorilla) on the one hand and the Asian great apes (the orangs) on the other. Physical appearance, it seems, is not always a sound guide to the evolutionary relationships that lie beneath the skin. So should the apes – or perhaps even just the African apes – be included in the club of 'moral beings' (those capable of holding moral views or being moral)?

One of the main reasons we are convinced that we should accord equality of rights to all humans is that we all share the same cognitive abilities from empathy to language. So the test might rest on the question of whether or not any of the other great apes share these traits with us.

So, do apes have language? The first attempts to teach languages to apes in the 1950s were notoriously unsuccessful, but that was because psychologists had tried to teach English to species that lacked the vocal apparatus to produce the sounds of human speech. They were palpably more successful when, setting verbal languages aside, they tried to teach them sign languages. So far the American deaf-and-dumb language, ASL, has been taught to several chimps, a gorilla and an orang, while languages that use arbitrary shapes on a computer keyboard to stand for words have been taught to nearly a dozen bonobos and chimpanzees.

By far the most successful of these has been the justifiably famous Kanzi, a bonobo or pygmy chimpanzee. Kanzi's ability to understand spoken English sentences, and reply using his keyboard, is now legendary. To be sure, neither Kanzi nor any of the other apes has language in the sense that you and I have it. In fact, their language skills are probably comparable to those of a three- or four-year-old human child, at best.

But, in one important respect, language is really only a clever means to an end. Of itself, it is merely a mechanism for transmitting knowledge from one individual to another. The real issue is surely the mental abilities that underlie language. So we are forced in the end to confront the thorny problem of exploring minds without the benefit of language.

So, what is it, then, that makes us human? The answer we are being driven inexorably towards has to do with the ability to understand the mind of another individual. As we saw earlier, recent work by developmental psychologists has suggested that human children lack this capacity (known as 'theory of mind') when they are born, but develop it quite suddenly at around four years of age. Prior to this, children do not realise that other individuals can hold a belief about the world that is different from their own. If they know that someone has eaten the sweets in the tin, then they assume that everyone knows that. But eventually they come to realise that others can hold beliefs which they know to be false.

The importance of having a theory of mind is that it opens the way to almost everything else that is human. It allows us to create literature, to invent religions and do science. It allows us to create propaganda, to be political and to produce advertising, for all these depend on the ability both to understand what is in another's mind and to manipulate the contents of that mind in order to change another individual's behaviour.

We now know that this unique ability, the very cornerstone in fact on which language itself depends, is not shared by all humans. Autistic people lack theory of mind: indeed, this is the essential defining characteristic of autism. Nonetheless, autistic people can be of normal, sometimes even supernormal, intelligence in other respects – remember the superhuman memory for numbers of Dustin Hoffman's character in *Rain Man*? What autists universally cannot do is handle social relationships, because they cannot think themselves into someone else's mind well enough to understand the subtle processes of human social interaction.

The substantive issue at this juncture is whether we humans are unique in respect of this ability. Despite the sometimes clever behaviour, even understanding, exhibited by your cat or dog, there is no evidence to suggest that any other species is able to think itself into another's mind. The only exception seems to be the great apes, but even they only do about as well as four-year-old children who are in the process of acquiring theory of mind.

But herein lies the quandary. For it seems that we share with great apes (even if only just) the special cognitive abilities like theory of mind that underpin our moral capacity and make us human, yet this is something that we do not share with all humans (infants, autists and the severely mentally handicapped seem to lack these capacities). But on the other hand, the genetics says we share more in common with these humans than with the great apes. So how should we decide who is a moral being and who is not?

No one would doubt the humanness of autistic people, any more than they would doubt the humanness of a one-year-old child. And no one would question either group's right to be treated to the full panoply of human rights. If we accept (as we should) that such individuals are eligible to belong to our community of equals, then we must ask ourselves how we should view those species that share the same set of cognitive properties even though they may not be quite so closely related to us as other humans are.

That said, it is one thing to say that we should feel an obligation to look after the interests of other species, and quite another to infer from this that these species have a human capacity to make moral judgements – although something like this did in fact happen in medieval times. A pig was once tried for murdering its master, whom it had gored to death. Duly condemned, it was executed for its heinous crime. We might find that bizarre now, but it is perhaps just another example of how easily we attribute these human-like capacities to have intentions to other species. The short answer is that there is no substantive evidence to suggest that any species other than humans have a moral sense. In this respect, maybe we are unique. That might be because having a moral sense actually requires more than second-order intentionality, and no species other than humans can aspire to that. It may be no accident that these high orders of intentionality are also required for full-blown religion in the sense in which we are familiar with it in humans, and that moral codes are invariably closely tied into religious beliefs. So let's finally turn to religion.

Chapter 22

How Evolution Found God

History tells us that not all the Victorians were impressed by Charles Darwin's ideas on evolution. They seemed to strike at the very heart of the biblical story of creation, and, what was perhaps worse, they challenged our exalted view of ourselves relative to the rest of creation. Wisely, Darwin chose to keep his opinions on the subject of religion to himself. And, following his lead, evolutionary biologists have, by and large, studiously continued to ignore God ever since, preferring to leave discussions of this rather contentious topic to sociologists and anthropologists.

But, in the last few years, God has finally come in from the cold and been placed under the evolutionary microscope. It is not clear what has triggered this interest, but a significant factor has probably been the growing realisation that religion *is* a real evolutionary puzzle – one that is intimately tied up with humans' sometimes disconcerting willingness both to behave prosocially (act altruistically towards those they never expect to see again) and, more puzzling still, to submit themselves to the community will, especially where religious belief is concerned. No self-respecting baboon or chimpanzee would ever willingly kow-tow to the good, the bad or the genuinely ugly in quite the way humans seem prepared to do.

We believe . . .

Religious belief is a real conundrum. In our everyday lives, most of us make at least some effort to check the truth of claims for ourselves. Yet when it comes to religion, it seems that we are most persuaded by stories that contradict the known laws of physics. As the experimental work of anthropologists Scott Atran and Pascal Boyer has convincingly demonstrated, humans seem to find tales of supernatural beings walking on water, raising the dead, passing through walls and foretelling and even influencing the future especially believable. But, at the same time, we expect our gods to have normal human feelings and emotions. We like our miracles, and those who perform them, to have just the right mix of otherworldliness and everyday humanness.

Why are we humans so willing to commit to beliefs we can never hope to verify? You might well think, along with that great paragon of philosophical common sense Karl Popper, that this question falls well outside the realm of scientific investigation. But evolutionary biologists have begun to challenge that convenient assumption. Given that religious behaviour seems to be universal among humans, and is often very costly, then it becomes increasingly difficult to duck the issue and write it off as froth on the evolutionary landscape. On the face of it, religious behaviour seems to be at odds with everything biologists hold dear. The reductionist view sees us as mere vehicles for our selfish genes – yet religions embrace charity to strangers, submission to the will of the community, and even martyrdom.

Even so, the biggest stumbling block for evolutionary biologists has been recognising that religion might have a functional advantage. If a biological trait has evolved, we want to know what use it is – and by that we mean how its possession makes an individual better adapted to survive and pass their genes on to the next generation. That's not always apparent where religion is concerned, especially where Franciscan charity or martyrdom are concerned. This apparent maladaptiveness of religion has prompted some evolutionary psychologists and cognitive anthropologists to conclude that religion is simply a functionless by-product of some more useful aspect of our cognition that is directly involved in fitness-maximising behaviour.

While it probably is true that religion parasitises cognitive mechanisms that evolved for some more general purpose, it does not follow that such behaviour is biologically non-functional or maladaptive. For one thing, the claim that something so costly in terms of the time and money spent on it, never mind the costs of martyrdom, is functionless is simply naïve: it is singularly unlikely that anything that costly could evolve even as a by-product of something else. Besides, humans just aren't that stupid. The problem really arises because most of those who now dabble in this area and promote the religion-is-maladaptive view are cognitive scientists and psychologists rather than evolutionary biologists: as a result, their understanding of evolution is, shall we say, at best challenged. They think only in terms of immediate benefits to the individual: I choose a mate, and benefit by siring offspring with them.

But for social species like primates in general, and humans in particular, this isn't always so. Multi-level selection processes are especially important for us because many of our solutions to the problems of survival and successful reproduction are social (we co-operate to achieve those ends more successfully), and social solutions require an intermediate step – making sure that the community pulls together. This is not to be confused with group selection – evolutionary biologists' big bête noire and unacceptable no-go area because it assumes that the benefit to the group is all that matters. Rather, this is to observe that some benefits to the individual come through group-level functions. That's a very different thing, and its implications have not been widely appreciated until very recently.

In recent years, evolutionary biologists including myself have come to realise that there are some important aspects of religion which do seem to have explicit benefits. In identifying these, we can start to pin down the origins of religion itself, leading us towards answers to two fundamental questions: why is religious belief so common, and when did it begin?

We can identify at least four ways in which religion might be of benefit in terms of evolutionary fitness. The first is to give sufficient explanatory structure to the universe to allow us to control it, perhaps through the intercession of a spirit world – religion as a form of primitive, albeit flawed science that allows us to predict and control the future better. The second is to make us feel better about life, or at least more resigned to its vagaries – Marx's 'opium of the people'. A third possibility is that religions provide and enforce some kind of moral code, so keeping social order. And finally, religion might bring a sense of communality, of group membership.

The first idea – religion as cosmic controller – seems highly plausible, given that many religious practices aim to cure diseases and foretell or influence the future. It was the view favoured by Freud. However, believing I can control the world is not the same thing as actually being able to control the world, and one might expect a species as smart as humans to figure out that it doesn't always work. So this proposal seems inadequate as an explanation for humans' apparent willingness to believe religious claims despite the evidence. Rather, I suspect that this benefit came about as a by-product once our ancestors had evolved religion for one of the other reasons – and thus had a big enough brain to figure out some metaphysical theories about the world.

The second hypothesis, Marx's opium, seems more promising. In fact, it turns out that religion really does make you feel better. Recent sociological studies have demonstrated that, compared to non-religious people, the actively religious are happier, live longer, suffer fewer physical and mental illnesses and recover faster from medical interventions such as surgery. All this, of course, is bad news for those of us who are not religious, but it might at least prompt us to ask why and how religion imparts its feel-good factor. I'll come back to that later.

The other two options are concerned with individuals benefiting from being part of a cohesive, supportive group. Moral codes play an obvious role in ensuring that group members keep singing from the same hymn sheet. Nevertheless, the sort of formalised moral codes preached and enforced by today's major religions are unlikely to provide much insight into the beginnings of religious belief. They are associated with the rise of the so-called doctrinal or world religions with their bureaucratic structures and the alliance between Church and State. Most people who study religion believe that the earliest religions were more like the shamanic religions found in traditional small-scale societies. These are quite individualistic, even though some individuals – shamans, medicine men, wise women and the like – are acknowledged as having special powers. Shamanic religions are religions of emotion not intellect, with the emphasis on religious experience rather than the imposition of codes of behaviour.

In my view, the real benefits of religion – and, as it happens, the explanation as to why religion makes you feel happier and healthier – have more to do with the fourth hypothesis. The idea that religion acts as a kind of glue that holds society together was in fact originally suggested by Émile Durkheim, one of the founding fathers of modern sociology, though he could say little about how or why this might be. A century later, we know a little more about how this works. Religions bond societies because they exploit a whole suite of rituals that are extremely good at triggering the release of endorphins in the brain. Endorphins come into their own when pain is modest but persistent – then they flood the brain, creating a mild 'high'.

This may be why religious rituals so often involve activities that are mildly stressful for the body – singing, dancing, repetitive swaying or bobbing movements, awkward postures like kneeling or the lotus position, counting beads – and occasionally even seriously painful activities like self-flagellation. Of course, religion is not the only way to get an endorphin fix. But perhaps that is why genuinely religious people often seem so happy: in a very real sense, they are getting their weekly fix. What's more, and here's the rub, endorphins also 'tune up' the immune system, and that probably explains why religious people are healthier.

Of course, you don't have to get your fix from religion. You can also get your high from jogging, pumping iron or many other forms of physical exercise. But religion seems to offer something more. When you experience an endorphin rush as part of a group, its effect seems to be ratcheted up massively. In particular, it makes you feel very positive towards other group members. Quite literally, it creates a sense of brotherhood and communality that doesn't seem to happen when you do the same thing on your own.

Thanks be to God

While this may explain the immediate advantage of religion, it does raise the question as to why we need it at all. The answer, I believe, goes back to the very nature of primate sociality and so takes us back to Dunbar's Number. Monkeys and apes live in an intensely social world in which group-level benefits are achieved through co-operation. In effect, primate social groups, unlike those of almost all other species, are implicit social contracts: individuals are obliged to accept that they must forgo some of their more immediate personal demands in the interests of keeping the group together. If you push your personal demands too far, you end up driving everyone else away, and so lose the benefits that the group provides in terms of protection against predators, defence of resources and so on.

The real problem that all such social contract systems face is the 'free rider' – those who take the benefits of sociality without paying their share of the costs. Primates need a powerful mechanism to counteract the natural tendency for individuals to free-ride whenever they are given the chance. Monkeys and apes do this through social grooming, an activity that creates trust, which in turn provides the basis for coalitions. Exactly how this works is far from clear, but, as we saw earlier, what we do know is that endorphins are a vital ingredient. Grooming and being groomed lead to the release of endorphins. Endorphins make individuals feel good, providing an immediate motivation to engage in the activity that ultimately bonds the group.

The trouble with grooming, however, is that it is a one-on-one activity, so it's very time-consuming. At some stage in our evolutionary past, our ancestors began to need to live in groups that were too large for social grooming to provide an effective glue. Such large groups would have been especially prone to exploitation by free riders. Our ancestors needed to come up with an alternative method of group bonding. In the past, I have suggested that gossip played this role, allowing individuals to perform an activity that provides a similar function to grooming but in small groups rather than one to one. But conversation lacks the physical contact of grooming that triggers the release of endorphins.

So what might have bridged the endorphin gap needed to bond these larger groups? Although laughter and music would have filled that gap, religion seems to have played a crucially important role in the later stages of human evolution. Religion seems to have been the third leg in the trilogy of mechanisms that supplemented grooming to make the later stages of human social evolution possible.

It is important to emphasise, however, that, if this account of the origins of religion is right, then religion began very much as a small-scale phenomenon. Perhaps early religious practices included something like the trance dances found in shamanistic-type religions today. The !Kung San of southern Africa, for example, seek to heal rifts in personal relationships within the community by using music and repetitive dance movement to trigger trance states. Many religions have practices such as chanting and fasting that invoke similar mental states: blinding light bursts within the head, the soul seems to become united with God and the mind has the experience of leaving the body and entering another (spirit) world. Doing this as a group seems to create an explosion of goodwill and love that welds the group together. It is easy to see how this sort of activity could have been extremely beneficial to our ancestors, uniting the group, discouraging free riders, and so increasing the chances that individuals would survive and reproduce more successfully.

Whence came the gods?

Religion is not just about ritual, it also has an important cognitive component – its theology. My suggestion is that the reason why religion has both ritual and theology is that the endorphin-based group-bonding effects of the rituals only work if everyone does them together. And this is where the theology comes in: it provides the stick-and-carrot that makes us all turn up regularly. But to be able to think about the nature of a divine being and its relationship with us, our ancestors needed to evolve sophisticated cognitive abilities that far exceed those found in any other animal species. And it is this aspect of the cognitive underpinnings of religion that provide us with an insight into the other question that has long remained unanswered: when did religion first evolve?

Our ancestors did not always have religion, yet many religious practices seem to have very ancient origins. So, when did religion first evolve? Archaeologists have long been fascinated by this question. But how do you recognise religion and religious practices when all you have is a few old bits of pottery? Being cautious folk (and having had their wrists slapped for idle speculation all too often in the past), archaeologists have perhaps inevitably defined the appearance of religion by uncontroversial evidence such as grave goods in burials: these at least unequivocally imply belief in an afterlife.

Although it has been claimed that the very earliest evidence of deliberate burials dates back as much as two hundred thousand years to the Neanderthals, the motivation for the kind of caching of bodies we find in this case is ambiguous. If we take grave goods as the only uncontroversial evidence for deliberate burials, then burials do not occur much before twenty-five thousand years ago. The oldest yet found is a child burial in what is now Portugal; the best known is an elaborate double burial of two children at Sungir outside Vladimir on the Russian steppes that dates to around twenty-two thousand years ago. Burials imply a sophisticated theology, so we can safely assume that these were preceded by a long phase of less sophisticated religious belief. But without evidence on the ground, can we realistically see any further back into the past than this?

Well, maybe there is another way to gain insight into the question. It comes from asking what kind of mind is required to hold religious beliefs. Take the statement: 'I believe that God wants . . .' To grasp this, an individual needs theory of mind. But we need more than this to build a religion.

Third-order intentionality allows me to state: 'I believe that God wants us to act with righteous intent.' At this level, we have personal religion. But if I am to persuade you to join me in this view, I have to add your mind state: I want *you* to believe that God wants us to act righteously. That's fourth-order intentionality, and it gives us social religion. Even now, you can accept the truth of my statement (that I truthfully believe this to be the case), but it doesn't commit you to anything. But add a fifth level (I want you to know that we both believe that God wants us to act righteously) and now, if you accept the validity of my claim, you also implicitly accept that you believe it too. Now we have what I call communal religion: together, we can invoke a spiritual force that obliges, perhaps even forces, us to behave in a certain way.

So, communal religion requires fifth-order intentionality, and this also happens to be the limit of most people's capacity. I think this is again no coincidence. The majority of human activities, from making tools to surviving the minefields of our complex social world, can probably be dealt with by the capacity for second- or third-order intentionality, yet the two extra layers beyond this undoubtedly come at some considerable neural expense.

Since evolution is frugal, there must be some good reason why we have them. The only plausible answer, so far as I can see, is religion. And that's where this line of reasoning can throw light on the origins of religious belief.

As we saw earlier, the level of intentionality a species can achieve seems to scale linearly with the volume of its frontal lobes. Perhaps we can use this relationship to work out the level of intentionality our extinct ancestors were capable of – provided you have a fossil skull from which you can measure the overall volume of the brain.

Plotting these values onto a graph, the evidence suggests that as early as two million years ago, *Homo erectus* would have aspired to third-order intentionality, perhaps allowing them to have personal beliefs about the world. Fourth-order intentionality – equating to social religion – appeared with archaic humans around five hundred thousand years ago. But the fifth order probably didn't appear much before the evolution of anatomically modern humans around two hundred thousand years ago – early enough to ensure that all living humans share this trait, but late enough to suggest that it was probably a unique adaptation. Interestingly, if we apply the social brain relationship to fossil hominids, it suggests that these same two key dates – five hundred thousand and two hundred thousand years ago – correspond to major up-surges in social group size, with the second of these corresponding to a fairly rapid shift from groups of around 120 to around 150 individuals that we find in modern humans.

Let me add one final caveat. All this does not justify the truth of religion as such. It simply offers an explanation as to why religion evolved in the human lineage – and only in the human lineage. Strictly speaking, I suppose that leaves open the possibility that the claims of religion, at least in some form, might be true: God might have chosen to reveal himself to humans at some particular moment in time, as some have argued. But I wouldn't find that a terribly convincing argument myself. Why then, and not earlier or later? And why only to our species, and not to any others? If there really is something transcendentally special about religion, it would seem to me an odd coincidence that it should appear just at the point both where the cognitive capacities to support it first evolve and where we find the glass ceiling in group size that needs both of these phenomena to break through. That said, true or false, religion does seem to work, at least on the intimate social scale. It does have benefits for the individual. But its real benefits seem to be in creating closely knit communities. It is only when religion is taken over by the state and becomes large-scale that problems arise. It seems that the psychological forces it can call on are so powerful as to be able to turn perfectly rational individuals into bigoted mobs. It is these psychological mechanisms that have been exploited down the ages by political elites in various attempts to subjugate the rest of the community.

Marx, it seems, was right after all. In his famous phrase, religion really is the opium of the people – in a much more literal sense than he probably ever imagined. But, equally, Durkheim seems to have been right in suggesting that religion played a key role in bonding small-scale societies. It evolved to make us toe the communal line, and it uses rituals to exploit the brain's own opiates to do that. The endorphin rush we get from all that singing and praying helps us to overcome the fractiousness of everyday human interactions and so gives us the crucial sense of belonging that welds together all traditional small-scale communities. But it seems that religion does its work best if it has a cognitive dimension to it – a reason for believing in what we do in the rituals. And here, in the magical bringing together of deep thought with what seems like no more than a base chemical trick, lies the impenetrable mystery of human relationships. In this respect, religion is just one of many archetypal examples of the way evolution has exploited and honed simple processes to create the extraordinary complexity of cognition and behaviour that makes us what we are. Evolution is truly a marvel, and it was Darwin's genius to recognise the processes that underpin it.

Index

Index

Index

Index

Index